RIDE LIKE A GAUCHO

SOPHIA ASHE

- Second Edition -

CONTENTS

CHARACTER LIST

The Family
Leonardo -- My host
Gisela -- Leonardo's wife
Carolina -- Their daughter
Eduardo -- Their son

Miguel -- Leonardo's brother, my host at the ranch
Patricia -- Miguel's wife
Rodrigo -- Their eldest son
Lucas -- Their youngest son

The *San Eduardo* Crew
Sergio -- The manager
Ricardo -- The hunt manager
Branzidi -- Hotel groundskeeper
Juli – The ranch cook

The Gauchos

Cabeza -- The foreman
Chango -- The groundskeeper
Chaque
El Tío
Enano
Gecko -- *'Postero'* for *Las Nutrias*
Gillo
Javier
Paisa
Pato
Pavo
Petaco -- The tractor man
Pirulo
Ruso
Tío Flaco -- *'Postero'* for *La Vigilancia*

MAP OF *'SAN EDUARDO'*

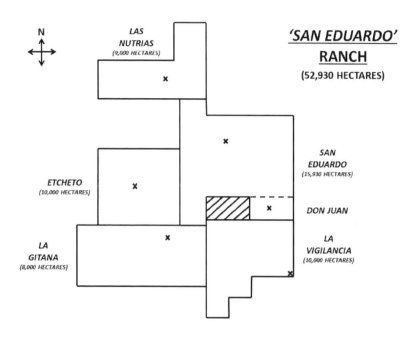

N

LAS
NUTRIAS
(9,000 HECTARES)

'SAN EDUARDO'
RANCH
(52,930 HECTARES)

SAN
EDUARDO
(15,930 HECTARES)

ETCHETO
(10,000 HECTARES)

DON JUAN

LA
GITANA
(8,000 HECTARES)

LA
VIGILANCIA
(10,000 HECTARES)

There are too many people to thank for their part in my travels, but this goes out to my parents and all my Argentinian friends

BYE-BYE BLIGHTY

When I graduated from university, I decided to do what a lot of graduates do: run away from the responsibilities of adulthood for as long as possible.

The completion of my academic programme hit me really hard. I loved my time at school and was devastated when that came to an end. Then the thought of going to university and looking after yourself entirely, deciding on which lectures to attend and cooking your own food seemed incredibly daunting - not to mention having to start the process of making new friends all over again. Anyway, I went on to also love my chapter at uni where I had a great time and made some amazing friends but before I knew, it those 3 years came to an end too and this time I was absolutely terrified by what was about to start. I was being released into the real world where I would actually have to try to be a responsible adult and the reality of being independent and focussing on a job for the rest of my life became very real. A lot of my friends had already secured jobs to go to in to from the moment they left. I, on the other hand, thought I'd do the complete opposite and ignore this problem for a little while longer but instead of enrolling in a Masters' programme as many unsteady graduates do, I did what the other proportion of graduates do and ran away to the other side of the world.

I make myself sound rather cowardly but my actions were somewhat encouraged by my parents, who told me that if I did not seize the opportunity to travel and explore at that stage, I probably never would do in the future. Having completed a degree to work in the agricultural and farming industry, it is tough to escape that world once you are in mud up to your neck in it, both metaphorically and literally speaking in this situation. Therefore for once I did as I was told, followed my instinct and planned a working holiday trip around the world for a few months. The beauty of working on farms is that you can chase the seasons in different countries whilst earning money. I was also eager to go out there to learn about the different farming systems adopted in different parts of the world; something I have a huge interest in. Originally, my plan was to go to Australia in October 2019 for the harvest season. I intended to spend 4 months there and travel on to Argentina for 6 weeks to fulfil my lifelong dream of cattle ranching by horse. Since the age of 3, I've been passionate about horses and a very keen rider. However, with the combination of losing my own horse and the long, demanding hours of farming, I had spent much of my latter years out of the saddle. Thus, the chance of combining both my passions – farming and horses – into one was just too good an opportunity to miss. Also, being a Spaniard, I thought it'd be fun to get in touch with my Latino side and try my hand at being a gaucho.

So there it was! After graduating, I left England at the end of September 2019 and got a one-way ticket to Brisbane. Annoyingly I had picked to go to Australia during their dreadful 2-year drought, so going for the harvest never worked out as there were basically no crops. Prior to leaving, I tried and tried to find somewhere through friends and family to assure guarantee that where I was going would be safe and suitable for a young lady travelling alone. This was one thing which made my family terribly nervous, especially since agriculture is still very much a man's domain. Fortunately, I had reassurance in the form of relatives and close friends residing in Oz, which meant that I wasn't totally alone if things didn't work out. Anyway, despite my parents going through my entire list of close friends trying to see if any of them would want to come along with me - with the months passing and too many undecided responses - it was set in stone that I would embark on my journey alone. This didn't really bother me at all. I was most content to travel alone because it allowed me to go about my business at my own pace without having to coordinate the trip with anyone else. I had total liberty to do everything I wanted to do, and that's how I liked it.

I was totally winging my trip to Australia since I left England with no job lined up, simply hoping to find something once I got there. I stayed a week in Noosa with my relatives, where I adjusted to the time difference and the heat whilst also enjoying the white sand beach and morning coffees along the turquoise river. While I was busy enjoying these moments, I was applying for jobs. Miraculously, on the whole, the outline of my plan worked and when a job came up on a cattle station, I bit the bullet and accepted the place on a 40,000-acre property with 2,000 cattle near a town called Winton, in the scorching Outback of Queensland. For this, I had to return to Brisbane and take a 22-hour bus ride, which was, without a doubt, one of the worst trips of my life. The only good thing was that the bus was practically empty, but other than that, I have no compliments whatsoever. Trying to sleep on a bus that frequently slammed its brakes on during the night to avoid hitting kangaroos meant that proper sleep was totally unobtainable. Luckily for me, I was on the bus with a girl of my age whom I had met in Brisbane. This meant I had someone to talk to because the trip would have otherwise been completely boring once we got into the Outback – not that it hadn't been before. The roads were straight and seemingly endless, cutting through a flat, barren and featureless landscape. It was in fact so boring and featureless that there were great big yellow road signs at the side of the road, which along the top read 'FATIGUE ZONE' in bold black letters. Below that would be a trivia question. Further down the road, a follow-up board containing the answer would be located, perhaps two kilometres or so away.

I stayed at the station for 2 months and, much to my relief, the family were kind. I was well looked after, meals were served at the same table as the bosses' and the accommodation was more than adequate. My bedroom was small but ideal, fitted with a desk, a fridge, an *en suite* bathroom and - most importantly, an air conditioning unit right above the bed. Thank goodness for the air-con as it made sleeping possible. I actually happened to read quite an amusing article from a scientific journal while I was out there. The article stated that 'in order to not affect your health and keep down running costs, air conditioning units should not run lower than 8 degrees Celsius below the temperature outside' and this made me laugh.

'Much good it'd do me having mine at 37°C!' I laughed to myself. It was so hot out there that if I came back from work early and showered too soon, the water would be too hot to withstand. What's even funnier is that whenever this happened, it was better to shower using only hot water, as the cold water (which came from a collection tank outside) would actually run hotter as it spent the day being boiled by the heat of the sun. This was just one of the things I learnt along the way - the hard way.

I was in good company on the Station with a crazy Dutch backpacker, Chanelle, who had also travelled to Oz for a working holiday. I hugely enjoyed her company and we quickly became good friends. Being so isolated and away from absolutely anything or anyone, it was a huge relief to have someone of my age whom I got on quite so well with. We'd laugh at just about anything and during any journey into Winton (over an hour and a half away), we blasted out music, never crossing more than five other vehicles on the highway. We made a trip out of this drive, no matter what we went into town for. Our bosses were really chilled, but the only strict rule was not to drive on the main road in the dark because of the 'roos', so we always made it back to the station in the daylight. On our day off, we would spend the day at the lovely, refreshing pool, eating ice cream and on a separate occasion I accompanied Chanelle to her dentist's appointment. The dentist was a travelling dentist, a bit like the Royal Flying Doctor, but instead of flying they travelled in what looked like a massive American RV-style motorhome. Chanelle was so excited about her appointment, which was totally odd to me, but she is a dental nurse so I guess it kind of made sense. When we arrived back at the station, the boss asked her how everything went during dinner time.

'Great!' she replied. 'The dentist said I have the best teeth in Winton!'

'Yeah, don't let that go to your head' he mumbled in his thick Aussie slur.

We also found a strange pleasure in documenting the lives of the cane toads that were always outside the house when we would walk back to our accommodation after dinner. We decided these cane toads were husband and wife since the bigger toad would plod along behind the smaller one and frequently get slapped by it, despite the difference in size. Later on, we were joined by another backpacker - this time a Norwegian fellow, a couple of years older than us, who was also good company. Between myself and Chanelle, the poor chap didn't stand a chance. There were times I even felt sorry for him as we would pick on him and wind him up. All in good humour though, of course. Thank goodness he was a calm, placid soul who took everything we threw at him, because the poor chap had to put up with being subjected to the back seat on our trips into town, whilst Chanelle and I would sing tunelessly at the top of our lungs to whatever would come on from our playlist. Usually, this would be Bon Jovi, Cher, or Shania Twain as a special treat for our Norwegian pal who absolutely loved *'Man! I feel like a woman'*... although in retrospect I reckon we killed that for him. The poor boy even had his birthday at the station, which meant he was woken up by our horrific singing and when he opened the door of his room to go for breakfast, he was blasted with confetti from party poppers. 'All part of the experience,' we assured him. He must have appreciated it though, or else he wouldn't have stayed in touch, so that's a good sign.

My time in the Outback was a real eye-opener and I experienced a huge list of things I never would have even dreamt of, including mustering cattle by air and going to the 'local' pub (which was actually 40km down a dirt track), at the end of almost every week for some pool and darts. We also went on Melbourne Cup day, which was quite an experience in itself seeing the pub jam-packed for the first time (seeing as whenever we went we were always the only ones) and full of Outbackers effortlessly putting away bottle after bottle of beer like there was no tomorrow. On a separate occasion, I also spent an afternoon in the passenger seat of one of the tractors being driven about by my boss's 7-year-old grandson whilst he stacked bales. At first, it started the other way around and the young boy sat with me, but his father said I should let him have a go should he ask to, which he did. Apprehensively, I swapped seats with the boy and was left speechless at how perfectly he operated the machine and stacked up the bales, despite having to operate the tractor stood up as he was unable to reach the pedals properly - even with the seat as far down as it could go. Despite all this, however, the memory that will forever stick with me is when I was chased by a 6-foot-long King Brown snake. I was out with my boss at the station and I had exited the truck to open a gate when suddenly, he started hooting the horn frantically and reversed the vehicle. I then saw what he was pointing at and without a second thought, ran away from this evil creature which was now only a few metres away and coming directly at me. I dived into the truck, slamming the door shut after me and ensuring it was definitely closed. The snake followed us as we continued reversing away until it totally disappeared, which was suspicious as there was physically nowhere it could have hidden, given that where we were was just mile after mile of flat bare earth. It was then that my boss concluded that it must have got up and under the truck, especially seeing as the snake's tracks in the dust just stopped without leading anywhere.

Despite the uninvited hitchhiker, my boss drove through the gate, stopped the vehicle and said, in typical Aussie fashion, 'I still want that shut'. I told him there was absolutely no way I would get out of the vehicle with a great big lethal snake underneath, but eventually, I lost the argument and after being told to simply 'step out far and bravely', I did so. As a precaution, when we got back to the yard, the truck was parked away from any of the buildings at the homestead. The following day, the snake's departure tracks could be seen in the sand. A few weeks later, when Chanelle was on her way to the house to prepare dinner, just before actually entering the house, she heard a thud. She turned around and saw - in a heap on the ground - less than 2 metres from where she stood, a huge brown snake that appeared to have fallen off the roof. A fraction of a second earlier and it would have landed on her head. Given its huge size, we all thought that perhaps after having hitch-hiked a ride into the courtyard from under the truck over a month ago, it was quite happy and decided to stay there, resurfacing from its slumber at that moment, only to return to hiding quite quickly upon being chased by the men with shovels.

I'd love to say this was my only experience with dangerous animals, but it wasn't. In fact, I knowingly came within a metre of a brown snake on three different occasions and almost let a scorpion into my room one night as it appeared suddenly out of a crevice in the concrete right outside my door when I opened it. I say knowingly, because God only knows how many times I got close to something lethal that I was totally oblivious to. The brown snakes were pretty much the exact same colour as the reddish-brown Winton earth and the clumps of dead, golden grass did nothing but help their camouflage. So did the gaping cracks in the ground, which looked like portals to the Underworld, created by the evaporation of every last molecule of moisture. It, therefore, seemed very fitting that they too should be riddled with overly venomous reptiles, because why not? Everything else seemed to be. Chanelle and I often discussed snakes and agreed that one thing about living in unexciting countries is that you don't constantly need to watch where you are putting your feet in order to prevent stepping on something deadly. Although out there, to some lesser extent, you did get used to being surrounded by things that wanted to kill you. One evening after work, we three backpackers were sat on the concrete outside our accommodation enjoying a nice cold Coke and a packet of crisps when a small scorpion ran by, inches away from my fingers. None of us even flinched. At night, while walking back to our rooms in the dark using the torches on our phones to see, we would invariably have to stop to give way to the foot-long centipedes which scuttered about in the darkness.

After my two months on the station, I had some travel time to see some of the beautiful sights in Australia. I was hugely fortunate to spend a weekend on Magnetic Island cuddling koalas with Chanelle before we then parted ways. After which, I spent a few weeks with family friends, which comprised of spending a night on their sailing boat on the edge of the Great Barrier Reef and watching turtles lay their eggs on the beach at night. Using a very faint torchlight, we would walk up the beach late at night looking for recent turtle tracks (which, to my amusement, look like the tracks from a tractor tyre) on the sand and follow them to see if we were lucky enough to find the turtle laying her eggs, which we were on a couple of occasions. We would be out for hours, watching the turtles lay and then follow them slowly back to the sea to watch them swim away into the great ocean.

Another great memory I hold is from when I visited a saltwater crocodile farm, not far away from where my friends lived. This was both terrifying and fascinating in equal measures. Being just the other side of a wire fence to Australia's hugest and most ruthless killers was very unsettling – but one could not escape from being captivated by the strange beauty of these modern dinosaurs as they shot out of the water, opening their immense jaws to grab the chicken carcase hanging from the guide's hand. Even if you're there to watch it hatch from its egg and spend time with it every day of its life, a 'salty' will never bond with a human. In fact, it will try to eat you at the first opportunity. They appear to be immune to any type of infection and disease and they do not bleed out. This is why salties can lose limbs in fights against other crocodiles and survive the injuries until they die of natural causes at 70 years of age – or more.

Throughout this down-time, I kept searching for a job. Eventually I found one, this time working with ostriches down south in Victoria for 6 weeks. On arriving down there, I was amazed at the difference in country and terrain between the station and Victoria. Winton was once an inland sea (and also known as Australia's Dinosaur Capital due to its quantity of fossils) and, as a result, is totally flat and stony with no trees apart from a few gums, or 'ironbarks' as they are known, along the veins of the dry creeks. All of this combined reminded me of photos of the landscape on Mars, whereas Victoria was totally different. There, instead of eucalyptus, *Leylandii* and other cypress trees dominated and the rolling fields decorated with remaining round bales from the harvest really meant some areas could fool you into believing you were back in England, or so I thought at least.

I was in the hotspot of the brutal bushfires which engulfed almost the entire country that year. On my flight from Queensland down to Victoria, the entire area around Sydney, or at least what I could see that wasn't covered in a sheet of smoke, was either burnt to black nothingness or ablaze. It was a heart breaking and terrifying sight. Victoria became an area riddled with fires but luckily I was safe from them the whole time. However, one night, a fork of dry lightning struck a stubble field on the neighbouring farm and set it alight. Luckily, the field was right next to the farmhouse and, having heard a tremendous crackle, the farmer went out to see what the noise was. Upon finding his field ablaze, he immediately and successfully worked on putting it out. The field was left with a small black scar, but thank goodness nothing more. This all happened just two minutes up the road, so I really did count my blessings. Had the farmer been away or fast asleep this could have been a whole different story, which would have included an emergency evacuation and the release of two thousand ostriches to run for their lives to avoid being roasted.

I found working with ostriches hard and repetitive. The days were long, non-stop on your feet all day and all of this was made even worse by the stupidity of these creatures. They say sheep are stupid and like to find interesting ways of killing themselves but from my experience, compared to ostriches, they deserve a Nobel Prize for intelligence and could perhaps be deemed one of the world's easiest-to-look- after animals. Fact - an ostrich's brain is smaller than its eye, which means when you stand next to a beast, perhaps 8 or 9 feet tall, the ratio between the body and brain strikes you as being unfathomably out of proportion. Over time, however, this fact was put into light and was indisputably accurate. They kept getting themselves caught in fences and spilt their water all over the place as soon as you filled it up. They would also get scared by everything. They drove me mad and all I could think was that a brain of that size was far too generous, because they could certainly make anyone believe they didn't possess one at all. One thing I couldn't stand was their inability to resist the urge to peck you whenever I moved close to them. I would drive the quad bike into the field to feed them and they would chase me around. As soon as I stopped, I would then be surrounded by twenty or thirty massive, inquisitive birds pecking away at both myself and the bike. Thankfully, nine out of ten times, these were curiosity pecks and did not actually hurt, but they were enough to disrupt whatever you were doing. The remaining one out of ten times, they did hurt, leaving a nasty bruise or a blood blister. I had to abandon trying to re-patch a hole in the fence because I was being pecked at as if I was a bucket of food and it was impossible to do anything. I also had an incident once where I had a glove in my pocket and one chick stuck its head through the fencing, pulled it out, and proudly ran off with it. This led to my poor glove being used as a tug o' war rope as they all went to attack it. Of course, I couldn't chase after the birds to retrieve it as they would have died of shock so I had to just wait for them to lose interest before entering the paddock, picking up my filthy, torn glove whilst being pecked at by forty or fifty irritating waist-height creatures along the way.

The chicks may have been incredibly aggravating with all their pecking and stupidity but luckily, they did not pose a threat. The adults, however, were another matter. Whilst checking a water supply, I was confronted by a huge angry female, about 6 or 7 feet tall, who didn't want me in her paddock. She charged at me with her wings out, hissing and bumping into me repeatedly with her chest in an attempt to knock me over. Luckily for me, she didn't actually manage to do so but she did plant one of her dinosaur feet in my thigh with a nasty forward kick, which I'm not ashamed to admit, sent me running to safety. Adult birds are no less inquisitive than chicks so, of course, I was surrounded by hundreds of these creatures, all eyeing me up to peck me. This meant that, in my frantic getaway, I had to weave my way through the birds, pushing them out of the way as I ran. I dived through a hole in the fence head first, as if by destiny, the exact same hole I was unable to repair previously due to the birds' constant pecking, and I tumbled to the ground in a truly *'Mission: Impossible'* manner. From the other side of the fence, this bastard bird stood over me, looking down at me with her beak and wings still open as I lay on the ground catching my breath, having just made a narrow getaway. My dignity was significantly damaged, but miraculously, all my leg got was a bruise from the kick. Ostriches may only have two toes on each of their huge size 18 feet, but you wouldn't like to know how sharp the claw at the end of each toe is. Much like a kangaroo, a successful kick could slice you open. As you're reading this, you might probably find the image of someone running away from an ostrich terribly amusing, just like everyone I told the story to. I expected some sort of sympathy from friends and family when I shared this experience, but instead, they all found it hilarious. Unanimously, they all said that when they think of ostriches, what springs to their mind is them dancing to Ponchielli in Disney's *'Fantasia'*, where they so sweetly bat their eyelids and prance around in their ballet shoes. Given the bruise she left on my thigh, I really wish that bird had been wearing ballet shoes.

It wasn't all doom and gloom with the fires and giant stupid birds, though. I'll never forget the laughs we had at dinners with my boss and his family when a silly game would be brought out after the meal, or spending New Year's Eve with the family and the other backpacker (another lovely Dutch girl), playing croquet on the lawn before dinner (my boss wasn't pleased that, thanks to me, we lost) and then watching the fireworks off the Sydney Harbour Bridge on television. It was also a chance to be very smug with everyone back home, calling them and wishing them a Happy New Year, although they would still have to wait another eleven hours.

I worked hard and learnt a lot from my time in Oz and as a graduate, from an agricultural point of view, I found it absolutely fascinating. I had always wondered how people in the Outback can live and work under such scorching heat for their whole lives and survive hours away from the nearest town, as well as live their lives starting work at 5:30am and finishing at 6pm (or later). This seemed to answer a great question of mine: Why are the Aussies alcoholics? Well, it's because there's bugger all else to do. As for the ostriches, the combination of their severe lack of brains and how incredibly (surprisingly) fragile they are made me constantly question how on earth they survive in the wild and haven't extinguished themselves. Regardless, I endured the scorching heat, lethal snakes and brainless Jurassic feather dusters, surviving everything wanting to kill me and even managed to find something to go on toast that wasn't

Vegemite. I completed the missions I had set out to conquer and might even add that these experiences made me tougher than old boots - except in Australia this just means you come out with dark, dry, cracked leathery skin as a result of the dust and the heat.

My flight to *Buenos Aires* was booked for mid-February. After saying goodbye to my ostrich bosses and the backpacker I worked with there, I spent a pleasant and tranquil few days exploring and relaxing in Melbourne with friends before my 18-hour flight to Argentina. I found travelling from one side of the world to the other very amusing as it was a true Time Lord feat, given that despite the length of the flight, I arrived in Argentina on the same day, only 4 hours after my departure from Australia - something I found rather baffling. I have always been one to brag about never having had jetlag before, but it is safe to say this journey was the exception and I had my sleep completely messed up for about 3 days. Anyhow, after a long flight and a battle during a two-hour layover in New Zealand trying to persuade the airport authorities to let me keep my jar of honey and my empty Ostrich egg (which remarkably made it back to England safely after travelling all the way inside my riding helmet), I finally arrived in *Buenos Aires* and I was absolutely thrilled. I stepped off the plane, took a deep breath and hesitated just a moment whilst I said to myself, 'wow, I am in South America!'

I had been waiting with such great excitement to get to Argentina for so long. With a massive smile on my face, I hurried through customs, got my luggage and swapped some dollars for a few *pesos*. I say a few *pesos* but, in reality, the *peso* is so weak against the dollar that I was handed over a wad of cash so thick that I felt as if I had just robbed a bank, since 1 US dollar at the time was almost 80 *pesos*. Being too much to carry in my purse, I stuffed it into my suitcase and carried only the amount I would need to pay for a taxi into the city.

BUENOS AIRES

February 2020. I had always rather fancied the thought of going to Australia, but visiting Argentina had never really crossed my mind but now that I was there, I was over the moon. My trip to South America was actually realised by my mother, who has Argentinian friends, so thanks to her I was received by her friend's best friend, whose ranch I would be spending my time on. I did not know these people and they did not know me. However, upon arriving in *Buenos Aires,* I stayed at their place in the city for a few days and they received me like an old friend. They had a son and a daughter, both a little older than me, and they were absolutely delightful. The daughter, Carolina, was very sweet and chatty. One night, after dinner, she took me out for a drink at one of the bars just next to their flat. It was heaving with people, mostly small groups of friends sitting together, drinking happily and chatting away at the picnic benches. As we approached the bar, she suggested I should try one of the Patagonia ales if I liked beer, and so I did. I had heard about them but never actually tried them before and certainly, it was one of the nicest beers I had ever tried.

As for the son, he was exactly how I'd imagined the Argentinians to be. Eduardo arrived home after work and greeted the housekeeper, Elena, loudly and cheerfully. Then he rushed through to where I was with his sister to introduce himself to me. He was a very tall and cheerful person. He spoke about his day loudly and enthusiastically. His face was always radiant with a smile and he would sing to himself whenever there was a break in conversation.

My time in *Buenos Aires* wasn't long as I was focussed on having all the days on the ranch I could get. Besides, the combination of travelling alone and not being much of a city person anyway means there is only so much exploring one person can do alone. I love visiting cities and seeing the sights but I get quickly bored of them as they're not really my element. Also, since I had arrived during the week everyone in the household was at work so I had no option but to go about my own business. Nonetheless, I made good use of my time in the city and explored it well, although the first day was a bit of a no-go due to the jetlag. Despite being exhausted from the journey I was wide awake at 3am and the fact that I was struggling to sleep the rest of the night meant I was up at a very unsociable hour in the afternoon which gave me little time to have a proper look around the city. The good thing was that where I was staying was just around the corner from the famous *Recoleta Cemetery*: a cemetery of grand mausoleums where Eva Perón is buried. I informed Elena that I would be popping out for a wander. After ensuring I had everyone's numbers in case of any issues and double-checking I knew where I was and how to get to the cemetery and back, I headed off. Elena was so lovely, always making sure everything was okay for me and repeatedly expressed how amazed she was at my bravery for being a young lady travelling so far from home all by myself. I spent a few hours wandering around the cemetery and the surrounding area, also popping into the *Basilica de Nuestra Señora del Pilar* right next door.

Before arriving, I was not sure of what to expect of *Buenos Aires*. I knew very little about Argentina before arriving and given what you hear about its poor economy, its occasional fiery politics and the warnings of pickpockets, I suppose I really didn't expect *Buenos Aires* to be what it was. I was very pleasantly surprised and I came to adore the city. Just this walk around the block on the cobbled streets, with the Tango dancers on the street and people sitting on the grass enjoying a few drinks together, instantly made me feel very relaxed and happy. The whole atmosphere was warm and welcoming, packed with happy people filled with the cheerful *Latino* spirit one expects.

I arrived in Argentina on a Monday and after speaking to the family, it had been arranged that I would be taken to the ranch by my host's son-in-law that Friday. I, therefore, had 2 days to have a good look around the city. On Wednesday, I tried my luck to do the touristy thing and see the city from a tour bus, but it was totally booked. Therefore, I resorted to seeing the nearby sights on foot, starting at the *Museo de Bellas Artes*, which had gorgeous artwork by some very famous painters, including El Greco, Rembrandt, Goya and Van Gogh. The night before, I had been chatting to the daughter who had been giving me pointers on nice places to see. She made a few recommendations for particularly beautiful streets and she said that the world-famous *Teatro Colón* Opera House was a must-see. As a lover of theatres and opera houses, I made visiting this place a priority. After the art gallery, I had a nice long walk down the *Avenida Alvear,* past the stunning buildings of the Alvear Palace Hotel, *Palacio Duhau* and the French Embassy on the corner joining onto the *9 de Julio* – the world's widest avenue. It was only after a short walk along this busy, tree-lined avenue with the symbolic

Buenos Aires obelisk in the distance that I eventually came across the outstanding opera house in all its grandeur. It is safe to say, buying a ticket for the guided tour was money well spent. I was left astounded by this building's history and beauty.

I had a long walk back, but when I finally arrived at the flat in the early evening (to be greeted by a rather flustered Elena who was beginning to get worried that I had got lost), I could not stop praising the beauty of the little bit of the city I had seen, especially the *Teatro Colón.* 'It would be amazing to watch a production there' I said to her, who was delighted that I was enjoying the city with so much interest. I had got back not too long before everyone else, so I took the chance to sit and speak with Elena for a while. She was from the *Corrientes* Province, which is all the way up in the north of Argentina in its more jungle-rich tropical region, bordered by Uruguay, Paraguay and Brazil.

The next day I had more luck with the tour bus and I spent the entire day going around the city, admiring the sights and stopping off at the *Caminito* in *La Boca*; the little street famous for its row of multi-coloured houses, and for a spot of lunch at the *Cafetería Tortoni,* just off the *Plaza de Mayo.* As a sucker for historical architecture, this square was yet another location I fell in love with. Home to the *Casa Rosada* (the house and office of the Argentine president, which, at least from the outside, makes 10 Downing Street look incredibly modest), the National Bank and the Cathedral, this square was simply beautiful and a lot of the city's architecture was reminiscent of buildings in France and Italy. Seeing how the time had passed me by, I used the tour bus as a taxi to get back to the flat but just before getting back I swung by a shop at the bottom of the road that I had passed the day before on my way to the *Teatro Colón.* It caught my eye as it sold beautiful traditional gaucho and Argentinian goods: horse tack, leather items and ponchos. I had a look around and I had my eye on a *'mate'*, which I thought would be a nice souvenir to take home. 'I'll buy it now', I said to myself, 'just in case I don't get the chance to drop in when I return to the city before flying back home'. This turned out to be a very good decision indeed. As I stumbled into the flat I was greeted by a slightly panicky Elena, who was getting worried about me as I had been out for so long. Later, I was introduced to my hosts, who up until this point, had been away working. It was obvious to see where the son got his height from when a very tall, slender gentleman came out of the living room to meet me. This was Leonardo, the friend of my mother's friend who had made this trip possible for me and from behind him appeared his wife, Gisela, who was thrilled to finally meet me. She pushed past Leonardo and rushed over to give me a huge hug. Like I said, as if I was an old family friend.

That evening, the family decided I must try an *asado* so they took me to a traditional-style restaurant which found itself located on a smaller back road in the city, not too far from where they lived. When we got out of the car, I could have easily found my way directly to the door of the restaurant just by following the mouth-watering smell of smoke and sizzling meat that lingered across the street. Upon opening the door, the restaurant was jam-packed with people eating to the right and to the left, where most European bars and restaurants would have a large drinks bar and till, there it was replaced by two massive *parrillas* over open fires, with the grills fully loaded up with the most amazing quantity of meat. I say meat, but of course it was all beef. In fact, if Argentinians say *carne* (meat), they mean beef. If the meat is anything other than beef, then it was given its proper name: lamb, pork, chicken - whatever, but if it's beef; *carne* with vegetables? Beef with vegetables. Stew of *carne?* Beef stew – and so on.

As we were shown to our table upstairs, I glanced over at what was sizzling away on the *parrillas* and I had it pointed out to me that what was cooking were the favourite cuts of beef: *tira de asado* - a rib strip (sometimes known as a Jacob's ladder beef rib), *matambre* and *vacio* - basically flank and skirt, and offal, including kidneys and sweetbreads.

'I hope you *really* like beef because you'll be eating plenty of it on the ranch', I was warned by my host.

It was delicious. I had heard all the rage about Argentinian beef but having never tried it I assumed it would be just as good as a grass-fed British steak... How wrong I was. The taste was absolutely divine; it had a special herby sweetness to it and the smoky essence from being cooked on the open fire on the *parrilla* was paired with the succulent juiciness of the tender meat (cooked rare). If the meat was to be as good as this out on the ranch, I was certain I would have no issue eating it every day. In the Outback, beef was on the menu for breakfast, lunch and dinner, no word of a lie. Sausages, steaks, stews, mince, corned beef sandwiches – everything. At first, I wasn't really bothered by this, but over time (and I'll be shot for saying this, I'm sure), I came to realise that I didn't really rate Australian beef. Of course, living surrounded by cattle at the station meant we eat our own, which is fine, except that the Outback has absolutely nothing growing on it so the beef is totally lean and in my own opinion, totally flavourless as a result. On top of that, everything was cooked well-done, so not only was it pretty tasteless but the steaks contained a moisture content to match that of the very ground the animal was born and raised on.

Over dinner, Leonardo and Gisela debriefed me a bit about the ranch but as they spoke, it was plain to see that they were not used to having girls working on what I was going to be doing and they were doing their best to make sure I was comfortable with what

I was about to embark on. They, as well as my mother's friend who had been to visit the ranch a few times, had repeatedly mentioned that their workers were good people, otherwise they would never have agreed to let me go out there to work with them. Of course, they were being polite and caring. Seeing me as a young lady, they apologised in advance for any foul language or rude phrases exchanged between the men. I said it would be fine. I'm used to it, and that much is true seeing as I have never worked on a farm with another female, aside from a few days' work I did with the Dutch backpacker with the ostriches. Gisela also reiterated that I was there as a guest, so I could do as much or as little work as I wanted with no obligation to do anything I didn't fancy taking part in.

'Please, anything you don't want to do or feel uncomfortable doing, you have no obligation to do. You can even contact us if you need,' added Leonardo.

'Thank you, I appreciate it but don't you worry. Anything they can do, I can do.' With a smile on my face as I said this, I think they were satisfied that I was going to be OK.

After a delicious dinner, we returned to the apartment and prepared for my departure to the ranch the following day, I packed a bag only with what I knew I would need and anxiously double and triple-checked that I had everything. I even decided to pack my riding helmet, which I had a feeling was going to be a long shot as I was probably not going to wear it at any stage. However, I thought it would be better to take it nonetheless.

BYE-BYE BA

I barely slept that night from the excitement and was up early, wide awake and ready to make my way to meet the son-in-law who was going to take me out to the ranch. I had said goodbye to everyone the previous night, since I would leave before everyone was up in order to make an early start on the long journey ahead. I let myself out and walked a couple of blocks down to meet the brother-in-law and two other members of the family outside their apartment. The ranch was located in the middle of *La Pampa*, which is where the geographical area of Patagonia commences. This meant our journey was going to be about 8 hours long, heading south-west for over 700km, not including stop-offs. Luckily, after having been in Australia, these long drives seemed like nothing to me and after spending 21 hours on the overnight bus from Brisbane to the Outback, I don't think there is any journey I could possibly endure in my life that could be quite so bad.

'Do you have everything you might need? There is of course a chance to go into town once you're there, but *Santa Rosa* is an hour and a half away from the ranch,' they warned me.

Yes, I was very well packed and being that distance away from the town was the same distance as the cattle station was from Winton in Australia, so this didn't bother me. The route from *Buenos Aires* to *Santa Rosa* was a long, straight road. Not a motorway, just a normal, wide road where, on either side almost the entire way, crops grew. We passed fields of maize, wheat, soy and peanuts, and as we approached towns (which were pretty few and far between, but nowhere near as bad as in Oz and, on the whole more civilised and lively), we passed agricultural machinery dealers with their great big forage harvesters and tractors, proudly advertising themselves on the roadside. As expected of a long journey, there were a couple of stops at service stations, mainly to refill the thermos flasks with hot water for the *en-route mates*. In the loos they even have designated bins for used *mate* leaves, demonstrating just how critical *mate* is to the Argentinians, far more than tea is to any Brit. Throughout the journey, the son of my host (who at the ranch was Leonardo's brother) sat in the front passenger seat and was in charge of preparing and handing out the *mate*. He would pour the water and have his drink, then fill it up with water and hand it on to the next person - and so this went on until either you no longer wanted any more or the leaves lost all their flavour.

Hours later, we finally entered *Santa Rosa*. *Santa Rosa*, as I was told, is the capital of *La Pampa* and is actually quite a big city, but on our way to the ranch we only went around it. Sadly, this of course meant I did not get to see any of it. However, with a month and a half on the ranch, I was sure I would be able to pop in at some stage to explore it. We wove our way through the smaller outer streets of the city before reaching a sort of ring road that took us past the *Santa Rosa* airport for domestic flights and along a gigantic water reservoir. We drove away, leaving the city behind us and as we did so, I looked over my shoulder to watch it disappear slowly into the distance, taking in the last glimpse of urbanisation before becoming totally immersed in wildlife. The road leading away from the city was long and straight, much like the rest of the journey was, but whereas before we crossed open, flat arable land, there the further we drove, the vegetation either side of us as far as the eye could see grew thicker with short, dark trees, more like big bushes. Every few kilometres or so we would pass entrances to ranches, marked by their large gates and wooden signs showing off all sorts of quirky and interesting names, like *Nuevo Mundo* (New World), *La Gallega* (The Galician), *El Venado* (The Deer) and *Vaca Muerta* (Dead Cow). I loved some of the names around there and even better still were the names of some of the nearby villages: *Carro Quemado* (Burnt Carriage), *Venado Tuerto* (One-eyed Deer) and the more traditional Indian names like *Chacharramendi* and *Quemú Quemú*. Whilst picking out all these interesting names as we drove on further out of *Santa Rosa*, a final *mate* of the journey was called for.

'Come on Sophia, will you try some now?' asked Lucas, Leonardo's nephew.

'Go on then!' Lucas passed me the *bombilla* and as I enthusiastically took it into my own hands, everyone in the car fell silent in suspense as I took my first sip of this new beverage.

It was very strange and not quite what I expected. To me anyway, it tastes slightly differently from how it smells, perhaps down to its odd, herby bitterness. I had been told about *mate's* bitterness. It accompanied a flavour that resembled a mixture between ordinary black tea and green tea, but with a much stronger, almost grassy flavour.

'I like it!' I exclaimed happily and proceeded to take more sips of this strange drink. There was a sigh of relief, followed by a small cheer. I did genuinely like it, but given how exalted *mate* is, I don't think I would have had the courage to tell them if didn't like it and most likely would have just played along with it in order not to hurt their feelings. Perhaps if I had insulted their revered drink, I would have been forced to finish the final leg of the journey on foot. Either way, I assured them that I did like it.

As we continued, *mate* now in my hand, along what was beginning to feel like an interminable road, my excitement was getting the best of me. Is it on the left? Is it on the right? How much longer now? I became increasingly impatient to finally arrive at this new place and see what it was like. As we drove, the bright orange sun slowly lowered itself behind the trees and moments before all the light of day had gone, the car slowed down and we turned left into a gate, which next to it had *'SAN EDUARDO'* in great wooden letters. I was beside myself with excitement but sadly, with the fast-fading light, it was soon pitch black and this very final bit of the journey across the ranch felt as if I was doing it blindfolded. At first it was annoying not to be able to look at my new surroundings, but at the same time it made it more exciting as I would wake up tomorrow to completely new sights.

After about half an hour of just driving from the gate on the road across the property, we arrived at the homestead at just gone 9pm. As we drove across the yard, warm yellow lights lit up the buildings and we pulled up alongside a small patio surrounded by a few buildings. There, I was greeted by Miguel, Leonardo's brother, and shown to my room, which was located in the corner of this patio area. As I opened the door, I was met by a lovely little place with a red brick floor, a small wood burner in the corner and an *en-suite* bathroom. There were two little single beds, each with its own bedside table and a tapestry on the wall at the bed heads. On one side of the window, deer antler coat hangers were fixed to the wall and on the other side were two obsolete gun mounts. Before building the hotel, which stood about a kilometre away down a track from the homestead, this was one of the rooms which would have been used to receive the hunters. So, it was aptly decorated for such, but was now only used for guests.

I freshened up before dinner and it seemed that a lot of the family had gone to the ranch to spend the weekend together, so I was introduced to cousins, daughters, nephews and grandchildren. It was a bit daunting being a total stranger amongst all these family members, but they were all lovely people and made me feel very welcome, especially Miguel and his lovely wife, Patricia. Of course, to celebrate the gathering, we ate none other than an *asado*, which I couldn't believe was even more delicious than what I had eaten in *Buenos Aires* the night before. Wow! If I was to be eating meat this good for 6 weeks, I was going to be in paradise. The room we ate in was its own little building just behind my room. The *'quincho'*, as it was called, was a small square building exclusively reserved for the family and guests. As soon as you entered, you were instantly struck by the ambience of this small room, which was the typical sort of rustic-style ranch dining area you could imagine or have seen in films: it was painted white inside and out and as you entered, there was a wooden bar on the right and behind it were the sink and fridge, with glasses, cups and biscuit tins on shelves above. Immediately in front of the door was the wooden dining table and chairs, and to the left were a couple of sofas and a tea table in front of the massive open chimney with its *parrilla*, which had a television beside it in the corner of the room - the only non-rustic item present. The floor was lined with terracotta tiles, the ceiling was beamed and thatched and the whole place was surrounded by big windows, which I imagined would give this place a lot of light during the day. Overlooking all this from above were mounted heads of trophy deer, black-buck and boar, which had been hunted on this very ranch.

The manager, Sergio, joined us for dinner so he could meet me. He was young, say early 30's, about my height, slim with black hair and a very gentle, friendly expression. He hesitantly sat down next to me, introduced himself and spoke in slow, steady Spanish - then looked very surprised when I replied to him fluently.

'Oh, what a relief! I didn't know you speak Spanish' he sighed and immediately looked more comfortable.

He seemed a delightful chap, friendly and polite, and outlined some of the work I'd be doing. He got me to follow him to where the workers eat their meals, just in case any of them were still around so he could introduce me. I followed him across the patio, not really able to see very much even with the lighting on the outside of the buildings, and followed him into an eating area where two men were seated at an empty table. They had obviously just finished eating and were chatting. As we entered, the conversation broke and they looked over at me with shy, inquisitive eyes. Our meeting was brief with very little exchange. I didn't get their names and all I got to say to them was '*Hola*' when Sergio introduced me. He explained that I was the 'English girl' that had come to gain experience on the ranch and that, to everyone's relief, I spoke Spanish fluently. The men didn't respond. Instead, they just politely nodded and smiled at us as we made our way out.

We returned to the *quincho,* where Sergio then handed me the item of ultimate importance: my own walkie-talkie.

'Have this with you at all times,' he said to me.

There was fragile internet at the homestead but absolutely no phone signal anywhere on the ranch. So, if anything were to happen, or I needed to communicate with someone whilst out working, the radios were all we had. It was a funny sort of arrangement I had concerning my work on the ranch. In Oz, I was an employee at both the station and the ostrich farm. On the ranch however, they had all the workers they needed, but they did not want to refuse me the opportunity to gain the experience I desperately wanted. So, as reiterated by Gisela and Leonardo when we went out for dinner in *Buenos Aires*, they said I would be greeted as a guest; fed and looked after and I could do whatever work I wanted without obligation. This seemed perfectly fair, although I knew I would be out on a horse working at every opportunity and throwing myself right in, doing a bit of everything there was to do.

'Tomorrow some of us are going horse riding in the morning. Would you like to join?' asked Miguel.

'Of course!' I replied without hesitation.

'Very well. I have asked Sergio to tell the workers to select nice, safe horses for you to ride. Generally the horses are good here but there is the odd fiery one. Tomorrow, he will also take you to pick your own saddle which you will use as your own whilst you're here.'

After an exquisite dinner and happy conversation, I politely excused myself and bid everyone goodnight. It had become very late and I could barely keep my eyes open from the combination of a long day of great excitement and my slowly fading jetlag. As I was about to leave, Miguel asked me if I had a torch, which I thought was an odd question to ask seeing as my room was right next door and there were a few lights on walls along the way to illuminate the path.

'Not as such, though I have one on my phone,' I replied.

'That's okay, I just wanted to check. Whilst most of the *casco* is well lit, some areas aren't. Although it's all fenced with gates and cattle grids, it hasn't been unheard of for a buffalo to stray through the boundary. They're angry creatures. You wouldn't want to cross paths with one, so just be careful.'

I cannot for a moment say that was the way I imagined that conversation to go, but I would be lying if I didn't say it brought a smile to my face. It just reminded me that I was in the middle of nowhere, totally surrounded by wildlife. One thing was being surrounded by wildlife in the form of snakes in the Outback, but this was an entirely different sentiment. In fact, it was an incredibly refreshing feeling. With that, I trundled over to my room, eyes open for any buffalo, got into bed and fell straight to sleep.

The next morning, I woke up extremely excited and threw apart my curtains with great anticipation. My room overlooked a grassy area where a small red brick chicken house stood about 15 metres away from my window. Beyond it was a field with a few sheep in it. I wandered across to the *quincho* for some breakfast and coffee but for some reason, despite being super excited, I was also so strangely nervous that I couldn't stomach the thought of eating. As I then walked to meet Sergio where he had described to me at dinner last night, I admired my pleasant new surroundings *en route*; all the buildings were similar to the *quincho* in appearance, in such a way that they were simple in structure, painted white and most of all, just had a ground floor. The *quincho* was the only building with a thatched roof, whereas all the others were of corrugated iron. As I walked under the covered walkway outside my room and past the family's house, I went along beside the butchery before reaching the block of apartments (for lack of a better word) where Sergio, Ricardo and the foreman lived, which was in view from my bedroom window beyond the chicken house, right near the fence of the sheep field. The entrance gate to the *casco* was just a little farther beyond their houses and I noticed then, in the daylight, that it was also painted white. Instead of standard gate posts, there were two small wagon wheels on either side of the redundant cattle grid.

The ground across the entire ranch was pure sand but at the homestead, there were nice, well-tended areas of *grama*, just like the bouncy, robust sort of grass commonly found in Mediterranean countries. There were no stone or gravel paths. The sand had been made easy to walk on, presumably by the traffic it had received over the years, which resulted in it becoming compacted and hardened. The only areas which were not sanded were the tiled stone patio, the covered walkway outside my room and a hardstanding of concrete outside the entrance to the butchery. The whole homestead lay beneath the shade from the canopies of tall, beautiful trees; mainly elms and poplars with the odd oak tree. In later days I came to discover that this great dark canopy proved to be a useful landmark as it stood out from the pale landscape and could be seen from miles away. It helped orientate me and stopped me from getting lost, since I'd at least be able to make my way back home.

As I approached his house, Sergio came out and took me to where I would pick my saddle, from the barn which faced my room across the little patio and was adjoined to the *matera*. The *matera* is where the workers sit, convene and eat their meals and, of course, as the name suggests, share their *mates*. A buffalo skull hung over the large, heavy metal doors and as they opened, I could see all sorts of saddles; from polo-type saddles, a Chilean saddle with a pommel like a Western saddle, to the typical Argentinian *recado*, which I wouldn't strictly classify as a saddle: they use a sponge as a saddle cloth, put the leather 'saddle' on top, which is then all tied down by the stirrups and girth (which are attached to the same piece of leather). Using an overgirth, you can then tie down as many cushions, blankets, or sheep-skins as you wish to make it as soft and comfortable as you like. Although I was keen to try the *recado*, I decided to opt for the Chilean saddle to start with, as I like the comfort of Western saddles and I felt something a bit more familiar would be better to re-accustom me to being back on a horse. When I picked it up, I was surprised at how light it was, having expected it to weigh like a Western saddle. Smiling proudly with my selected saddle in my arms, I followed Sergio out of the barn.

'I'll now take you to meet our sub-foreman, whom I've asked to look after you. Of course, you can come and see me at any time but, since you'll be out riding and working with the boys, I thought it'd be logical to have someone keep an eye on you whilst you're out and about. You'll get to know everyone in due course but at the weekends, only a handful stay to keep an eye on things and because it's a national holiday, the rest won't be back until Tuesday.'

'Thank you. How many workers do you have here then?' I asked.

'In total? About 20. We've got designated people across all areas: we've got the horsemen, the tractor driver, the groundskeeper, the fencers and the plumbers who keep an eye on the water systems and the windmills' (like in Australia, they use windmills to extract water from bore-holes). 'Then we also have a couple at different *puestos* across the property and Ricardo, who is in charge of the hunting side.'

Wow! I had never been anywhere with such a huge team. To begin with, I was worried I wouldn't remember everyone, so perhaps it was a good thing I had arrived at the weekend to get to know just a handful of people first, before all the others returned.

After selecting my saddle, I hung it briefly on a fence whilst Sergio took me to his office to show me a map of the ranch (and printed out my own for me to carry around). He described everything about the ranch to me. As a newcomer and never having experienced something like this before (not even the station in

Australia was this complex as it was quite simply fenced into 4-sections of 10,000 acres), it all sounded rather confusing. All in good time, however, I would get to know it, see it for myself and hopefully, bit by bit, it would all begin to make a little more sense. The ranch, *San Eduardo'*, was 53,000 hectares. To put this into perspective, that's twice the area of Birmingham, which is a mere 26,780 hectares. Alternatively, since everyone nowadays seems to like comparing everything to the size of a football pitch, then that makes the ranch just over 100,911 football pitches big. *'San Eduardo'* was the ranch's conventional name (for lack of a better word), as it also had a native Indian name which translates to 'sweet water', probably due to the number of lagoons dotted around the place.

This massive place was home to almost 9,000 cattle; predominantly red and black Angus with Herefords, but also a few dairy crosses, Shorthorns and *Criollos* (a traditional Argentinian breed descended from Iberian stock imported many, many years ago) thrown into the mix. By and large, the cattle out there were smaller and stockier than what you might be used to seeing in England. They were fat, healthy and hardy, despite living a totally wild life on pasture with limited human intervention, aside from being regularly moved around to ensure they didn't eat the ground bare. The main property (34,930 hectares) was split into 4-sections: *San Eduardo, Las Nutrias, Don Juan* and *Etcheto*. The main *casco* (homestead) was located in the section of *San Eduardo* and that's where I was located with the majority of the workers. Then, there were areas with *puestos*: a *puesto* (marked by 'x's on the map at the beginning of the book) is basically a post at the ranch, so a little house in a certain area that an individual (*postero*) or a couple of people would live in and they would be responsible for looking after that area, its cattle and ensuring all water sources are working properly. *Las Nutrias* was looked after by a small family, whilst

Etcheto and *Don Juan* had old *puestos* that nobody lived in anymore. To make up for the rest of the land, there were also two rented properties south of *San Eduardo*: *La Vigilancia* (The Watch) – 10,000 hectares looked after by just one *postero* alone and *La Gitana* (The Gypsy) – 8,000 hectares that had been newly leased, where two of the team members would eventually move to. The sections were then divided into smaller lots, where each individual 'field' varied from anywhere between 200 hectares and 1,100 hectares. For instance, there was *Etcheto* 1 to 16, *Don Juan* 1 to 5 and *San Eduardo* that went from 1 to 49, despite a section that was actually classified as *Las Nutrias*. It was all logically mapped out, but knowing each lot did take some getting used to, especially when you'd hear something like 'we are moving cattle from 4 to 32.' This sounds like an incredibly long way, when in fact they're neighbouring lots and it means passing the cattle through a single gateway. Fortunately, the ranch was fenced into lots because as you venture further south, deeper into Patagonia, some of the ranches are twice the size of this one or more - and there isn't a single fence in sight. It's so arid and bare that the cattle and sheep must be free to roam in search of adequate grazing, which must be one hell of a job when it comes to locating them and herding them into pens.

The terrain was very diverse. As well as being a working cattle ranch, it is also run as a game reserve for the hunting of red deer, water buffalo, black-buck and wild boar, hence all the heads, antlers and skulls on walls around the place. The ground was pure sand; it was like walking and riding on a beach and *Las Nutrias, La Vigilancia, Don Juan* and *La Gitana* are generally what they call *monte* (which is what I saw from the road on the way there), which is thick with prickly acacias (*caldén*), carobs and other terribly spiny vegetation. Meanwhile, *Etcheto* was what they called *pampa*, which was the total opposite: open terrain with the most breath-taking scenery of rolling sand dunes and lagoons. After all, the word *'pampa'* itself comes from the language of the *Quechua* Indians, meaning 'a space without limits', a word which, of course, was assigned as the name of this province and many *Pampeanos* (people from *La Pampa*) will call this homeland of theirs *'mi Pampa infinita'* ('my infinite Pampa'). The *San Eduardo* section sat in the middle of these two zones and was a mix between *monte* and *pampa*.

We left the office. I picked up my saddle and we walked towards the horse lines and saddlery, which were just beyond the butchery, heading out in the direction of the main entrance to the *casco*. The saddlery was the only building that was slightly different from the others. Instead of being washed concrete, it was made of red bricks and the white paint had faded over the years, giving it a pale pink sort of colour. The windows and doors were green (like on the other buildings) and there was a corrugated iron-covered area out in front, where on wooden rails sat various *recados* and bridles hung from metal columns. Next to this, out from under the cover of the iron roof, was a wooden horse line with a water trough at the end of it where 4 or 5 horses were tied up. They were beautiful and it was reassuring to see them stood there with immaculate, unblemished silky coats which glistened in the dappled morning sun coming through the trees. Amongst them stood a young gaucho who looked to be in his early 30's and was about the same height as me, with thick black hair and a black stubbly beard. His hat drooped over the side of his face and he stood hogging one of the horse's manes with a pair of shears.

'Sophia, this is Pato, the sub-foreman. I've asked him to be the person to keep an eye on you.'

Now, everyone – every single person - on the ranch had a nickname that they were seldom referred to by anything else other than. Some of them had pretty basic nicknames, whilst others were very strange. Some were introduced to me by their real names, but this made it very difficult for me to recognise whom the others were talking about over the radio or in conversation, especially seeing as they very, very rarely referred to one another by their actual names. It is safe to say that a few weeks in, I was entirely using the nickname system and to some lesser extent, I had even been issued my own, too. However, seeing as at this stage I wasn't aware of the whole nickname malarkey and the fact that Pato means 'duck' (as in the bird), I was visibly confused. Unlike some of the others who were occasionally called by their real names, Pato never *ever* was. In fact, it took me two weeks to discover what his real name was. In Argentina they are far more tactile, a bit like on the continent in Europe, and when I arrived in *Buenos Aires* I quickly learnt that you don't shake hands when you meet somebody. So, wanting to make a good, friendly first impression and also having taken the lack of handshakes into account, figuring it was simply the done thing and the polite thing to do, I greeted this gaucho the same way I had been greeted by everyone I had met so far and he was a bit perplexed when I stepped forward and cheek-kissed him twice. The poor man politely went along with it as if nothing had happened (and Sergio didn't seem to find my greeting to him odd, either) but upon noticing Pato's expression become even shier and watching his cheeks slightly redden, I wondered if what I had just done was wrong and felt rather uncomfortable. *'Good start'* I thought to myself, internally crippling whilst trying to maintain a confident outlook.

The horses for the family members were mostly ready and tacked up. So, as an excuse to brush away this awkward first salutation, Pato kindly asked me to leave my saddle on the rails and I went with him to catch my horse from the round pen, which was located directly in front of some of the workers' accommodation, between the entrance track to the *casco* and the track that leads you to the hotel where the hunters stay.

Whilst walking over to the pen, I was able to see even more of the homestead. I couldn't stop myself from grinning at how quintessentially 'ranchy' it looked. It had such an incredible charm to it due to how traditional it all was, which really made you feel like you'd stepped back in time. I was overwhelmed with happiness, one reason being how obsessed I was with *'Calamity Jane'* when I was younger. At 7 years old, after watching the film for the first time, I wanted to be like her, and why not? Riding across open landscapes while living in a close-knit community and having an amazing voice as a bonus seemed like a perfectly reasonable life goal to me. The way the homestead was set out, with the many different small blocks of buildings around the place, the rustic post and rail fences and handmade wooden signs, it could have easily been a film set for a small village. Given the number of people there, it really did have a village-like community feel to it. Horses, cattle, a traditional ranch and a village feel – I was going to be a gaucha Calamity Jane! With two-thirds of the way already there, I had better get my vocals warmed up.

Within the round pen stood my mount, a finely built chestnut gelding with a noticeably tranquil character and a broken white stripe down his gentle face.

'Do the horses have names here, or do you identify them by number or other means?' I asked Pato. With so many horses on the ranch, I assumed they would just 'know' which horse is which, but to my surprise I was told otherwise.

'No, they all have their own names. Generally, anyway. Some of them are identified by some particular feature, such as their colour or a defect or something.'

'Does this one have a name?' I asked about the chestnut.

'Yes, this is Moncho. Each worker has their own horses, too. You'll meet Moncho's rider when everyone gets back on Tuesday. You will be lent the nice horses the boys have to spare.'

I saddled up and bit by bit, the family arrived at the horse lines, all happy and chatting away together, greeting one another with a cheerful '*buen* día!' and excitedly looking at the horses tied up, hoping to find their favourite. Eventually, everyone either picked or was assigned their horse and we set off. Most of us were on horseback, but some of the family with the younger members were in a little cart, pulled by a heavy grey draft horse driven by Pato. Despite having a sheepskin, and perhaps because I hadn't been on a horse properly for over a year, I found the saddle hard and uncomfortable, with the stirrups and knots of the girth sore against my inner calves. Despite this, my happiness numbed the discomfort pretty quickly. I was thrilled to be back on a horse, observing this new place with all its sand, bushes and birds through Moncho's golden ears. We had a little canter, which after so long without riding, felt even more liberating and fun than ever, with the sun beaming down and something totally new to see would appear with every stride. During my mood of excitement, my head was miles away and when I finally came back, I noticed I was going far quicker than the others and had overshot many of them. In order to look less desperate and hoping terribly that they didn't think I was trying to show off, I pulled Moncho back and paid more attention to keeping up with the pace of the others.

After a while, we dismounted and tied the horses up to a large *caldén* tree. We stopped for some *mate* and cake, which were brought along by those who came along in the cart. I noticed that this field we were in didn't have grass and *olivillo* (a silvery plant with olive-tree-shaped leaves, hence its name), which covered the majority of the ranch, but that it was, in fact, alfalfa. Afterwards, I was told that there were four lots of alfalfa sown across the ranch, an incentive that was developed by Miguel's grandfather, who bought the property and was considered a pioneer of farming in the Pampas. The vegetation there, although diverse, was largely of poor quality. Thus, finishing cattle takes time. However, by having alfalfa, this extra protein and high-quality forage helps calves get fat much quicker and is what aided the grandfather in obtaining his huge successes during his very prosperous farming career throughout the 1900's. His achievements were phenomenal and I thoroughly enjoyed hearing the stories about this truly intelligent man. In fact, Juan Alberto Harriet became something of an icon for me because of his determination and agricultural achievements. I recall Miguel telling me over lunch one day that his grandfather was once the world's largest alfalfa grower. Thanks to this, he was able to produce 240 kilos of meat per hectare, whereas other producers were only averaging 100 kilos. This in turn led him to achieve an historical moment in 1919, when he sold 5,000 finished steers to the market in just a day. The meat was sold and shipped to Europe in response to a plea for help to aid the post-War food shortage. However, all these achievements were made to look small in comparison to the journey of how he started.

At just 15 or 16 years old, Harriet decided to move away from sheep production and start a business in cattle. However, there was a shortage of cattle and prices were high, so with complete ingenuity (and perhaps a touch of madness), he decided to head all the way down south, deep into Patagonia. There, he planned to buy 1,500 hardy native *Criollo* cattle from an ancient Indian tribe - somewhere in the foothills of the Andes in the *Chubut* Province. The trouble was, in those days, the railways did not extend very far and at the time, the most southerly line did not go much further beyond the north of *La* Pampa. So, with a group of friends, some helping hands along the way and a large string of strong, fit horses, he undertook a 1,200km journey by horse, negotiated with the Indians, and collected his 1,500 cattle, which he herded up to the most southern railway station and got them to his rented plot of land back in the province of *Buenos Aires*. The journey took 3 months and he lost around 300 cattle along the way. Apart from this, it was such a huge success that he undertook the same journey again a year or so later. Juan Alberto Harriet was since deemed to be the 'Pioneer of the Pampas' and even had a biography with this title written about him. They had a copy of this book over at the hotel, which Patricia lent me and when I wasn't able to finish it, I was left one as a gift by Gisela, in my suitcase I had left back in *Buenos Aires*.

We were out for a few hours and after an enjoyable ride, the family and I went to the hotel for lunch. The hotel was only a year old and it had the most exceptional views across the *salitral* (salt marsh). Surprisingly, there were a couple of salt marshes on the ranch. I learnt that these are formed as a result of strong heat after periods of rain which draws up water from deep in the ground via evaporation, which in turn, causes the natural salts in the ground to dry and crystallise at the surface. Only certain patches are susceptible to this, as it depends on how deep in the ground the water is maintained. The area enclosing the hotel was a reserve where nobody was allowed to hunt, so it was thriving with wildlife. By simply peering down the binoculars to get a closer look at the salt marsh there were deer, black-buck, buffalo and rheas everywhere. That was the beauty of the hotel's location - because the animals would gather around the *salitral* due to the minerals, so all sorts of species would be visible all together at any one time.

The family were very accepting of me, but I felt somewhat bad intruding on their activities. So, when the opportunity came up to do my first little bit of cattle work one afternoon after lunch, I eagerly jumped at the chance. It was there that I met Gecko (not his real name, of course) and his brother-in-law, who lived in the *puesto* at *Las Nutrias*. When I was introduced to Gecko, I recognised him as one of the two men who were sat in the *matera* the night I arrived and, with a very cheerful and friendly *'hola,'* I greeted him, with no kisses this time, in order to avoid embarrassment. Since the others had gone to town for the weekend and it was his turn to stay at the ranch, he had come to help up at the main homestead for the weekend. We saddled up and I was introduced to a new horse called Zaino Panzón, which translates to 'Potbellied Dark Bay'. This, I guess, is what Pato meant when he said some horses are named after certain elements or features that identify them.

Gecko, his brother-in-law and I set off in silence to locate any calves which had escaped the roundup during the week. As there were trees to hide under and the massive size of each lot, you'll inevitably leave some cattle behind. The ride was quiet, as I think we were all rather unsure of what to say to one another. As far as I had been told, the gauchos had been informed that 'an English girl is coming for a few weeks to ride and do some cattle work'. However, since the culture over there is quite different from England's, I reckoned they were unsure as to what sort of 'work' I was going to do. They were probably unsure as to how good my riding was too, seeing as controlled English riding with two hands and quiet transitions is very different to asking a horse to gallop from a standstill whilst shouting at cattle and directing the horse with one hand whilst frantically flapping the other. Having played a bit of polo, this way of riding wasn't an entirely new thing to me but of course, they didn't know that. They all seemed quite surprised when I spoke to them and a few of them commented on how good my Spanish was. The fact that Sergio and Miguel were surprised over the exact

41

same reason made me wonder just how much information they had been given about me prior to my arrival. It seemed like not a lot, but somehow I rather liked this as it meant I could tell them a bit about myself without them having heard it all from someone else beforehand. It must have also been a huge relief for all involved, simply knowing I would be able to communicate with them.

Either way, we rode largely without speaking, but I quite happily rode along with Gecko and his young brother-in-law, taking in the sights and spotting the different herds of buffalo, until eventually, we came across a group of calves that had been left behind. We began pushing them in the direction they needed to go, but being very naughty (and perhaps combined with me not doing the right thing), a few turned away and one spun around and ran in the opposite direction completely. It then didn't help that Gecko's horse decided not to cooperate and stopped dead in his tracks, bucking quite viciously, refusing to move on. I watched anxiously, hoping he wasn't going to be thrown off and get hurt but eventually, Gecko won the battle and led his brother-in-law away as they both tried to keep the rebel group together. Meanwhile, I figured I was more useful in trying to get the one calf back that had totally separated herself off. I spun the horse around and despite being a bit rusty and already hurting from the hard saddle, I managed to gallop after this young heifer, turn her around and get her back to the group all by myself. When I was reunited with the two chaps, they were quiet and didn't say anything for some time. Not until Gecko turned his head and, whilst peering out from under his hat, quietly said,

'...You can *ride*.'

I felt myself go bright red whilst I thanked him for what was, to me, the ultimate compliment. From video clips of gauchos working cattle and seeing first-hand some of the world's top polo players (Argentinian, of course), I've never seen anyone ride like the gauchos. Even a family friend from back home said that, on a trip to Argentina years ago, he got to ride with the gauchos and that, until doing so, he always considered himself to be a pretty good rider. So, to have one of them compliment me quite so early on filled me with the utmost pride and happiness. As they turned around and walked on, I gleefully leant forward over Zaino Panzón and patted his strong, dark neck jovially.

With very limited further conversation, we returned to the *casco* and unsaddled, but this lack of exchange didn't bother me. It was all very new to me and simply my presence there and working with me as a tagalong was clearly very new to them. It would take time to find stuff to chat about and I felt as if they were unsure as to what sort of stuff they could ask this stranger. All in good time, everything would come together. From their expressions and behaviour alone, I could tell they were kind people. Plus, this gave me the chance to slowly break into speaking Spanish again. At home, I will often speak to my mother in Spanish, but having spent almost 5 months in Australia speaking only English prior to arriving in Argentina meant my fluency was slow. Despite this, luckily everyone seemed to understand me, which frankly is more than I can say about myself with the gauchos. Their accents, combined with the speed at which they spoke, were certainly going to take some getting used to, but this was all part of becoming accustomed to the new environment.

THE WORK BEGINS

I had a rough idea of the work I would be doing at the ranch, but didn't know exactly what it was going to entail. As far as I was aware, I was most interested in spending my days on a horse with the cattle, which thankfully is what the largest portion of the work involved. Despite the huge size of the ranch, the dry climate and arid conditions of *La Pampa* means the grazing of the stock has to be very carefully monitored. One needs to regularly move the stock from one lot to another to ensure the vegetation does not become decimated beyond full regeneration. It's a fine line between not only ensuring the cattle have enough to eat, but also keeping the cattle in one place as long as possible to give other areas the chance to regrow, whilst making sure they're moved in time to allow the grasses and herbs to regenerate. Attack the grazing too hard and without rain (which falls rarely there), the land will not replenish. To give you an idea, the 9,000 cattle across 54,000 hectares means each animal has 6 hectares, or almost 15 acres. In England, on the other hand, the lush grass enables farmers to achieve stocking rates of around 1 cow per acre, which is roughly two cattle per hectare. It was Sergio's job to scout the ranch and observe the conditions. He would then come back and tell us which herds needed to be moved and to where. Moving cattle wasn't all I would be doing, though. I knew it was weaning season, so there was a lot of work that involved calves, which is always fun. To be honest there was such a huge variety of stuff to do that by the end I had branded, ear tagged, castrated and vaccinated animals, amongst plenty more.

That morning, the family were preparing to head off to spend the day and night at the cabin by one of the lagoons. They asked if I would like to join them, but feeling as if they had devoted enough of their attention to me already, I declined their kind offer. Besides, Gecko had told me the day before that there were bits that needed to be done, so if I wanted to help out, I could. I figured that, apart from being desperate to start the work, it would be a good idea to first get the hang of working with this small group and get to know them before plunging into work with a bigger group of people I had yet to meet.

So, we saddled up again. I met Cabeza, the foreman, that morning in the *matera*. He was older than Pato and Gecko and he also had a black, stubbly beard with a snip of hair down his chin, something I found was quite a typical style. He seemed nice, but being in a high position, he had a rather direct, no-nonsense attitude about him. Plus, he too was probably a bit unsure about me, but regardless he offered me one of his horses to ride - a black gelding called Patonegro. I ended up using this horse a few times whilst on the ranch and I absolutely loved him. He had a terribly uncomfortable canter, but he was so kind and sure-footed that I felt very safe on him. He was square and strongly built, a bit like a cob, but not fat or overly stocky and with finer legs – a handsome, gentle face and absolutely nothing seemed to faze him. Like most of the horses there, he wasn't very big; I'd say the average height of the stock horses was about 15 hands high but strangely, once I got on, I didn't seem to notice the size. Usually I'm not a fan of ponies and small horses, largely because being someone tall on a small horse means your balance becomes compromised and often, particularly at speed, if the horse goes one way, you go the other. Thankfully, this didn't happen though.

Gecko, his brother-in-law and another gaucho, Javier (surprise surprise, not his real name) and I rode off to yet again check for any stray calves and cattle in a different lot. Javier was very friendly and I recognised him as the other man who was sat in the *matera* with Gecko the night I arrived. He very much confused me though, because when he introduced himself to me he approached and greeted me with two kisses, so I just had to assume that there was no right or wrong, but purely a matter of waiting and seeing which sort of greeting worked with each person. He was very chatty and thanks to him, we all got talking. They started asking me stuff about my riding and farming, what brought me to Argentina and it seemed they were rather fascinated by me, a foreigner, who had appeared on the ranch to work alongside them, with the same passion for the countryside and the outdoors as they had. It was very sweet really, and this topic is what appeared to bring Gecko out of his shell a little more, although his brother-in-law still remained very quiet. We had a gentle ride and a nice chat and right at the very end, when heading back towards the homestead, I kicked my feet out of the stirrups and let my legs hang down. Returning to the *casco* from the direction we were coming from meant we would ride past the enclosure where some deer were contained. I had totally forgotten that a member of the family was going to be in there with Ricardo culling some old does and just as we were alongside them on the other side of the fence, the rifle fired, which spooked the horses. Luckily, good old Patonegro just shied slightly to the side and simply by swapping the reins to my other hand and giving them a gentle pull, he composed himself instantly. Gecko, who was just ahead of me due to his horse leaping forwards in reaction to the shot, turned around, I guessed to check if I was OK. He saw I was fine and looked at my feet, which were hanging down out of the stirrups. He turned to Javier.

'This one can ride, eh' he said, bluntly and to the point, and I felt redness return to my cheeks.

By lunchtime, the family had left for the lagoon so I asked the boys if I could join them for lunch in the *matera*. They gladly said yes and I rather shyly joined them for a nice *asado* with salad, all prepared by the groundskeeper. At the weekend, when the ranch cook went home the boys would take turns to prepare meals for everyone. The groundskeeper was yet another friendly man, tall and slim, with long black sideburns and a cheerful face that revealed a missing tooth when he smiled. As he came out with a tray fully loaded with beef *tira de asado*, he introduced himself as Santiago. The others called him Santi or Chango and since Santiago is a perfectly normal name, I assumed it was his name and that Santi and Chango were his nicknames. How wrong I was. Turns out almost 2 months later, I found out his real name, which was something *totally different* and that he was called Santiago because he comes from the province of *Santiago del Estero* in the north of the country. For the first few weeks I didn't really have much to talk about with Chango because, apart from at meals, we didn't really happen to cross paths. Plus, he was one of the chaps I found harder to understand because of his accent and how quickly he spoke, whilst also throwing in slang words I had never heard of into the mix. Anyhow, we'd often find ourselves sitting near one another at meals, so bit by bit we chatted more and soon got on really well. He was one of the quieter members of the team and a very good soul. He worked hard and never gave any trouble (the boys often teased one another non-stop), but whenever he did open his mouth to speak, it'd be hilarious.

'After lunch we have to vaccinate and castrate calves down in the feedlot. Will you come and help us?' asked Pato.

As if he had to ask, I very (over)enthusiastically told them to count on me being there.

'Great! Meet us back here at 4 and we will head over,' said Javier. 'Hope you're good at lassoing.'

Lunch was usually at around 12 and to avoid working in the heat, they had time off until about 4pm. Of course, this meant most of them would go off and have their *siesta*. The whole time I was there, I think I only slept three times as I hate sleeping during the day, but it was a good time to ring friends and family so I made the most of it and took the time to put my feet up and sit in my nice, cool room.

Although it was called the feedlot, it was nothing like the huge, horrid, barren pens filled with hundreds and thousands of cattle that you see in shocking photographs of American intensive farming. It was far from that. In essence, it was six different pens together in a block, with a handling system at the centre of it. Each pen had a long feeder down the middle for the concentrate feed (the calves were fed twice a day), a bale of alfalfa and, of course, a large water trough. There was also a length of shaded area, for the calves to hide from the sun or shelter from the rain. The first time I went down there, there were a couple of hundred calves spread out across the pens, but the stocking density of the feedlot fluctuated between 0 and 100 calves per pen when it was very busy during the weaning season, which was to be in the next few weeks. However, the pens would not stay full for long as the calves would be loaded into trucks and taken away to another property, also owned by the family. In essence, the only reason it was called the feedlot was because it was the only place on the entire ranch that had feed silos and the only time when concentrate feed was used. Apart from the calves briefly penned up there, all other creatures were nourished by the flora and roughage of the land.

I found that the way they castrated and vaccinated the calves went something like this: they had the calves in one pen and they would release 5 or 6 at a time into another smaller pen where we were all stood. The boys lined up side by side along the middle with their lassos and as a bull calf ran past, the ropes were thrown at it. Once it was down, the other chaps would quickly run over and hold the calf still whilst someone castrated, vaccinated and sprayed it with antiseptic. The whole event to them was just fun and, I suppose, when you're dealing with hundreds of animals at a time, you've got to find a way to make the work fun and entertaining, even though all that running back and forth just made it more tiring - for me, anyway. Gecko, Pato, Javier and Gecko's brother-in-law had a great time running around lassoing whilst I generally just followed them with all the sprays and vaccines. I did have a go at lassoing though (I couldn't resist) and over the course of two days I managed to lasso 7 calves in about 30 attempts, which is an embarrassing 23% success rate. Not amazing but it was something, and I was proud of myself for it and it felt good to hear the gauchos whooping and cheering each time I did manage to lasso one. As it turned out, it was here that I had my real icebreaker with Gecko, because I accused him of cheating since he would always place himself before me in the line and his lasso would sometimes knock mine out of the way.

Ranches produce some of the most natural meat you can get. Their pastures are not maintained or enhanced with chemical sprays and fertilisers, which means they are abundantly diverse in wild flora and fauna. But, this does mean that they have to be carefully managed since growth can't just be encouraged by throwing on nitrogen. On top of that, medicine use is practically non-existent but vaccination is necessary to stay on top of diseases. At specific times of year, both calves and adult cattle need jabbing. In *La Pampa*, they have to monitor leptospirosis, brucellosis, anthrax and foot & mouth to maintain herd health across the country but aside from vaccinations, no other medical treatments occur. When you have a large population of cattle living so extensively, you're not seeing all your stock every day and you can't just walk a lame cow to the yard to trim her feet and give her a shot of antibiotic. Like it or hate it, this more natural, wild lifestyle does therefore put to task the survival of the fittest and where other countries will cull in order to develop their herds or flocks, out there, nature will do it for you. It's brutal, but it's nature. Out there, you can't really be interfering and meddling with it.

That evening, Pato had to go and sort out a water pump somewhere and he asked if I'd like to join him so I could see more of the ranch, bit by bit. We drove for a few minutes before reaching the pump house, which sat by a large water reservoir marked by a tall windmill. Serving as little use whilst Pato played with the pump to try and get it working, I climbed the bank leading up to the water tank to get a little more height. Unfortunately, the sun had already set so I did not get to see it go down, but I wasn't too upset as I knew I had plenty of time to see some amazing skies. Instead, I watched the impending darkness creep over us and as the *caldenes* became silhouetted before disappearing into the blackness entirely, I gazed up, watching the stars quickly beginning to glow brighter and brighter in the jet-black sky. Within minutes, the sky was totally speckled with stars and as I stood gazing up, I saw a string of satellites and I couldn't contain my emotions when I saw various shooting stars suddenly shoot across the sky at once.

'Oh!' I gasped, then turned over to look at Pato, slightly embarrassed, wondering what he would think of me getting so excited at something he probably saw quite frequently out there.

'Shooting star?' he asked.

'Yes, a few of them. It's magical out here,' I replied quietly, 'although I'm sure you must be used to seeing these sorts of things.'

'Yes, we do see them quite a lot but they never fail to astound me and you never really get used to seeing them. Each time is just as special and it reminds me how blessed I am to live out here.' He spoke peacefully and full of emotion.

On the drive back to the *casco*, Pato told me a bit about the ranch and it was obvious he was the one to know just about everything about it.

'I grew up here.' he commented. 'My father worked here and we lived at the old *puesto* in *Don Juan*. I've been here since I was about 3 years old and when I was old enough to work I became an official member of the team. I adore this place and know it like the back of my hand.'

'Wow, you've been here a long time. That's incredible.'

'Yes,' he laughed, 'although nowhere near as long as Chaque, whom you'll meet on Tuesday. He was born at the old *puesto* in *Etcheto*, which is now unused, making his family the last to live in the place.'

The next day, the majority of the family were to head back to the city after their weekend at the ranch. I met with them at breakfast to say goodbye and thanked them for including me in their activities. They wished me luck and enjoyment during my time at *San Eduardo*. They set off, but Miguel's family stayed and he encouraged his 15-year-old son, Lucas, to join in with the cattle roundup that was taking place that morning.

It was Tuesday morning and as we tacked up the horses, all these cars arrived full of people: it was the workers, all returning from the village for their week of work. I was allocated Patonegro again, but this time Pato asked if I would be happy to use a *recado* instead of the Chilean saddle because, being unaccustomed to them, the other saddles can cause sores on the withers of the horses. Obviously I said yes. I didn't want to injure any of the horses and I was actually glad to swap away from the hard Chilean saddle, which felt like I was sitting on a tree trunk. At first, I found the *recado* was a nightmare to tack up. Layer after layer – sponge, leather, leather, girth, cushion - I watched as Cabeza tacked up his horse for me and I paid attention to how it was done so I would know how to do it next time... but I knew deep down that there was no way I would remember how exactly everything went. Saddling up a *recado* was like putting together a club sandwich in a very specific order of ingredients. Although it's not necessarily the order in which everything is put on (that bit is fairly logical), the difficult bit was lining up all the layers, ensuring each one was evenly placed on the back of the horse. You didn't want the sponges or cloths to be over to one side and the girth strap had to be perfectly centred, else your stirrups would be of different lengths.

When I got on, with so many layers of cushioning this saddle felt incredibly wide and I was worried it was going to pop my hips out. I felt like the little girls in the Thelwell cartoons who because of their fat ponies, their short legs stuck out horizontally. However, it was with an instant sigh of relief when I sat in it and it was as soft as a sofa. No wonder the gauchos can spend all day in the saddle without getting sore! Western saddles have the reputation for being the most comfortable saddles (and don't get me wrong, they are quite comfortable), but they didn't even come close to this. Due to the various components of the *recado*, it's also a much, much lighter saddle and with all its cushions and sponges it's also incredibly soft and comfortable for the horses. I asked a few of the gauchos about back problems in the horses and they all said not once had they ever experienced a horse with issues, almost exclusively down to the comfort of the *recado*.

I was so excited - my first cattle drive! To be on the safe side, not knowing what exactly I would be expected to do or how wild it was going to be, I put on my riding helmet. The gauchos politely said nothing, but they did all look at me in my helmet with a funny look. They of course use none other than their *boinas* – the stereotypical Argentinian hat (also known as Basque hats), or on particularly hot sunny days, some would wear a broad-brimmed fedora-type felt or straw hat. Lucas and I set off with Pato and Cabeza, and over time the others (who were at this moment unknown to me as they'd just arrived from the village) caught up with us. As we walked along, I paid attention to how they rode, so I could emulate it. Patonegro didn't seem to be bothered by my riding style whatsoever, but having been trained to be ridden in a certain way meant that if I could ride like a gaucho, I would get the best performance out of the horse. To my benefit, having never attempted any proper dressage meant adopting some of these gaucho mannerisms wasn't too difficult. First, the reins: they hung loose and had no contact on the horse's mouth whatsoever, something every English riding instructor would call 'washing lines', which is an absolute no-no. These horses are steered by relying on shifts in the rider's balance and pressure from the reins on the side of the neck, whereas English riding teaches you to maintain an even balance on the horse's back and your hands stay in one position. Secondly, stirrup length and rider's position: they rode with their stirrups very long, to the extent that some of them rode with their toes pointed downwards and in the saddle, they sat in a far more relaxed way; not slouched, but not upright and rigid, either. In general, it was a freer and less regimented way of riding, which after having spent my more recent years playing polo and hacking out, was quite easy to get to grips with. Although, I thought I rode with long stirrups, but I couldn't manage having them *that* long and out of habit I always forced my heels down.

When we entered the field about 40 minutes later, Pato gave instructions to Lucas and me on what to do. We had 200 bulls to round up and take to the pens at the feedlot before moving them on to new pastures after lunch. Not being a part of the usual team, Lucas and I quietly worked the cattle whilst the gauchos shouted and made all sorts of funny noises to get the beasts moving. These noises included an assortment of shouts and whooping, along with more obscure sounds which resembled the laughter of evil clowns or quick, short bursts of *'ai'*, shrieked in time to their horse's canter which made it sound like they were sat on a pin which pricked them in unmentionable places with every stride. The bulls were relaxed and seemed, for the most part, pretty docile, which meant we were able to slowly walk them to the feedlot back at the *casco* without much hassle. I must confess, I'm not entirely certain my helmet was needed, but it's better to be safe than sorry.

'Come on Sofi, you must shout!' yelled Pato, doing his very best to encourage me to make some noise but I felt too embarrassed to make these noises yet. It didn't help either that I spent too much time laughing at the sounds the others were making. Back in England, it's all done so calmly and quietly. You try to do everything with as little fuss and shouting as possible but out there, that way of working simply wouldn't work. For a start, the animals wouldn't hear you coming so none of them would bother to get up and start moving. Plus, you can't just shake a bucket of food and get them all to follow you.

After penning up the bulls in the feedlot, we unsaddled at the *casco*, washed down the horses and released them in the round pen, ready for after lunch. Lucas and I headed back to the house, where we would have lunch with his parents. When we entered the courtyard area, I saw this short little lady wandering about looking busy. I remembered one of the things Gisela had said to me in the city:

'Juli is the cook and cleaner. She will look after you, do your washing and does all the cooking. You'll recognise her when you see her, not just because she's the only woman at the *casco* but because she's very small. Oh, and she's Dominican, so you may not understand her accent... Don't worry, we don't either.'

I walked over and introduced myself to the busy lady. She was very sweet, permanently smiling and just as Gisela had said, it was no joke that she really was very small (she came up to my chest height). Juli pointed out that her kitchen was right next to my room so in case I ever needed anything, I knew where to look. Apart from Gecko's wife, who lived with him at *Las Nutrias,* Juli and I were the only women on the ranch. In agriculture, most of the workers on farms are men. In fact, on all the farms I've worked on (up until the ostrich farm), I've been the only female actually working on the farm. There might be ladies in the office, but not once have I worked with another girl. Therefore, I wasn't daunted in the slightest about working with an all-male team or even being the only woman present at the ranch. In fact, it has become so normal to me that it hadn't even crossed my mind that I might have been the only woman there. Besides, I suppose I have always been a bit of a tomboy. For some reason, all of my close friends at university were boys; when we ever went to the pub or out to dinner, it would always be a bunch of lads... and me. Anyway, despite this, it was nice to have some female company and Juli and I quickly became good friends and stood up for one another against the teasing from the gauchos. In fact, we would support one another, join forces and hurl some abuse back, which was always fun.

After a very delicious lunch with Miguel and Patricia, I asked Lucas if he would come back out riding, to which he said no. This meant it was time to be brave and venture out with the gauchos alone. After the *siesta*, we grabbed our horses from the pens again and saddled up... well, everyone else did. I gave it my best shot at tacking up the *recado* but being my first time, I was making a real pig's ear of it. As I tried over and over to figure it out and get everything properly aligned, I saw a young, very slim chap dressed in perfectly typical gaucho attire, standing next to his horse in the distance, subtly watching me repeatedly struggle and fail. Eventually, my faffing and struggling must have pained him so much that he came over and, whilst keeping his head down, he corrected my *recado* whilst exchanging very few words. I think he was trying to explain some things to me, but shyly avoiding eye contact and keeping his head down whilst speaking in a quiet mumble, I struggled to understand much of what he was muttering.

'Make sure your girth is further back, not right by the elbow like with an English saddle' was about the only thing I understood.

'Thank you, I'll remember that for next time.' He nodded, then slipped away. 'What's your name?' I called out after him.

'Ruso.'

'Nice to meet you,' and to this, he looked up and gave a quick smile.

After the short exchange, we mounted up and headed to the feedlot. This time Cabeza didn't join us but I now knew the names of the other two chaps: Ruso and Enano. Enano and I still hadn't spoken yet, but I had heard his name being shouted by the others. He looked quite young, certainly one of the younger workers on the ranch, and he was short with jet-black hair and went about everywhere smiling. What struck me about him was that he was wearing a t-shirt with the sleeves cut off, which revealed strong, muscular arms that were so tanned that they would make any sun lover incredibly jealous.

As we entered the feedlot, Ruso dismounted, opened the gate to the bulls and handed me his *rebenque* (whip), telling me to go in on Patonegro and push the bulls out. I can't say I wasn't nervous; not because of the bulls but because I didn't want to make some silly mistakes or fail to accomplish the task in front of these unfamiliar people and make myself look foolish. Regardless, I braved it into the pen riding Patonegro up along the fence and then started pushing these bulls out of the gate. Without too much fuss I succeeded and Pato and Enano drove the bulls away whilst Ruso and I pushed the trailing bulls onwards. We walked the bulls out for a while until Pato and I turned back to head elsewhere, leaving Enano and Ruso to finish taking the bulls away. They would meet us later, Pato told me, as we needed to check another lot for any cattle left behind.

By the time we reached the next lot the others had caught up and I was horrified to see that this next area we needed to scout was thick with *caldenes* and other prickly bushes.

'You just head up here in a straight line until you reach the fence on the other side. Keep an eye out for any cattle as you go, OK?' clarified Pato.

Oh my God, I was sure to get lost in so much vegetation in such a big place but nonetheless, I braved it into the prickly jungle and plodded along. I felt bad for Patonegro, who was trudging through all these thorns, but he really didn't seem to care. I looked out in search of any stray cattle but could see nothing across the sea of infinite thorns. Ahead of me I heard a twig snap and got excited, thinking I'd found a stray but was instead crossed by a group of deer which, come to think of it, was a far more exciting sight. This place was heaving with wildlife and every time you left the homestead, you would be guaranteed to see something, no matter how long or short you'd be out. It was amazing.

My black horse and I plodded on and on and came across nothing until we hit an electric fence. I stopped and listened but couldn't hear anyone at all so I picked up my radio and called Pato.

'I've reached an electric and in the distance there's a silo on my right.'

'Okay, that's fine. Stay there, we're on our way up.'

So, there I stayed until I saw someone emerge from the bushes. It was Ruso, but instead of coming up perpendicular to the electric as I had done, he was riding up alongside it.

'Oh... I went pretty far off course then I see,' I said to him.

'It's fine. You'll get used to it. Follow me and we'll wait for the other two.'

He gave his horse a nudge and effortlessly galloped off, so of course I followed but struggled to keep up. I decided against wearing my riding helmet, partly because of the funny looks I received, but also because of the heat and the sun. So, I swapped to my 'cowboy hat', which I ended up fighting throughout the entire gallop so it wouldn't fall off, despite the chin strap being done up so tightly it was almost choking me. Meanwhile, I noticed that no matter how hard they galloped, the gauchos never lost their *boinas*, which was pretty incredible seeing as this 'hat' essentially sat on the top of their heads like a pancake with nothing in particular to keep it in place. Some of the *boinas* had a leather sweatband that helped them stay on a bit better, whereas some of the others had nothing at all - not even elastic - yet they would barely ever part from their user's head.

We only waited a few moments before Pato and Enano joined us at the silo and we then headed home. It was a beautiful, warm evening and we rode alongside one another, gradually chatting more and more. Enano, for some reason unknown to us cantered off ahead after only a short while, leaving Pato, Ruso and me behind. I can't remember how it happened, but the three of us ended up talking about silly things which had us all in stitches. I, at least, was laughing until my stomach hurt. I still hadn't exchanged a word properly with Enano yet and since he charged off, it was a while until I finally got to speak to him.

Pato was easy to speak to. Having grown up in *Don Juan* while his father ran the *puesto* back when it was still lived in, I loved hearing all the stories he had to tell of *San Eduardo*. I enjoyed hearing him speak of his horses with such affection. He also had Indian ancestry, which I was fascinated to hear about, with all its traditions. In fact, his aunt was famous in the surrounding villages for hand-making ponchos in a particular way with a traditional loom, making her apparently one of the few people left to make woven goods in this unique way.

Ruso I came to learn was by no means short of words. He was just a bit shy in my first few days and, like the others, probably just unsure of how to react to this strange foreigner working with them. However, come to know him, he was totally mad and provided us with good laughs, and we got on well. It was on this ride home with the laughs, the setting sun and the odd sighting of a buffalo in the reserve that I knew from that moment that my time there was going to be very special.

NICKNAMES & HORSES

After a few days of working with the gauchos and sharing meals with them, I finally met everybody. I feel like I should dedicate a short chapter entirely to who was who: the nicknames of the workers and some of the horses. Behind some of these names, for people and horses alike, there were interesting back-stories and some rather amusing origins to all these bizarre and wonderful nicknames.

As Sergio had told me on my first day, unlike on English farms where everyone pitches in doing various jobs, on the ranch they had enough labour to assign to people to specific roles.

Firstly, you had the horse and stockmen: Pato, Ruso, Enano, Pirulo, Pavo, El Tío, Gillo and Javier.
Pato got his name when he was a young boy, because his neighbour saw him splashing about in a puddle and belly-flopping into it. Hence the name, 'duck'.

Ruso was called 'Ruso' because when he arrived at the ranch, his arms and face were very tanned but his body was totally pale. Hence, they said he looked like a *ruso*, which means Russian.

Enano means 'dwarf' but can also be used to describe something very small. Despite not being the smallest worker at the ranch, it would appear that Enano he was sufficiently small enough to receive this nickname. In addition, he was one of the younger members of the team, along with Pirulo and Gecko.

Pirulo looked after the feedlot calves whilst I was there, but was also a stock worker. He was called Pirulo because there's an Argentinian ventriloquist who has a puppet called Pirulo, which he apparently looked like.

There seems to be an avian theme in the nicknames because there was also Pavo, which means 'turkey'. However, it's also used to describe someone of a dippy sort of character. Needless to say, although he was a very sweet chap who never gave any trouble and was always a target for teasing, he was quite 'pavo'.

El Tío, which translates to 'The Uncle', was Pato's actual uncle but was called uncle by everyone else.

Gillo was simply known by his surname and was Ruso's older brother, even though you wouldn't believe it when you saw them; Gillo was a laid-back, tranquil soul with a relaxed way of talking whereas Ruso was quite simply bonkers and spoke by murmuring because his mouth couldn't keep up with his brain.

As for Javier, his real name was misunderstood by someone at the beginning and he became stuck with the wrong name.

Then there were the others: Cabeza, Chango, Petaco, Tío Flaco, Chaque, Gecko and Brandizi.

Santiago/Santi/Chango was the groundkeeper, in charge of keeping the homestead tidy and the lawns mowed. He also looked after the small flock of sheep, a few pigs and the hens. His nickname, as I stated previously, was because of where he came from.

Cabeza was the foreman and his nickname means 'head', which simply came about because he had a big head.

Petaco/Pepo was the tractor driver and was named after a famous *jineteada* rider who is *actually* a dwarf. Because Petaco was quite small, he was named after the rider. A *jineteada* is the gaucho version of a bucking bronco rodeo. I'll talk in detail about it later, as it's quite a feat of bravery.

Tío Flaco, which means 'Slim Uncle', was the *postero* at *La Vigilancia*. He lived there alone and looked after all the cattle on the 10,000 hectares single-handedly at the age of 63. He was another one of Pato's uncles, also called Uncle by the rest of the team.

Brandizi was another one known by his surname and he was the groundskeeper of the hotel and maintained the fishing lagoons. Then there's Gecko, who got this nickname because it rhymes with the nickname of his real name... so it is essentially a nick-nickname (it's all very confusing).

The last, but certainly not the least, is Chaque (pronounced cha-ke). I never found out where his nickname came from but the word '*chaque*' was a versatile one, used by the boys as an expression if something fell to the floor: '*Chaque*, I dropped my hat,' or indeed, that 'something' could also be a rider. So, if someone fell off their horse everyone would shout '*chaque!*' and laugh. It was also used to express surprise, for example: 'Diego caught his wife sleeping with the neighbour.' 'Really? *Chaque!*'

Even the fencers and the chaps in charge of the windmills had nicknames. It seemed that when you arrived at *San Eduardo,* you couldn't stay for long before receiving your own name. Gradually over time, I acquired a few myself. Usually I was just called Sofi but I had also been granted La Flaca (the thin one), Rusa (nothing to do with Russians or being pale. It's the Argie word for my brown/blonde hair), Pequeña (small one – Javier's nickname for me because I was the youngest person on the ranch), La Inglesa (the English girl), La Española (the Spanish girl), La Loca (the crazy one) and La Gallega.

Technically, this last one means 'The Galician' but despite telling them time and time again that I wasn't from Galicia they said that being Spanish meant I must be Galician, which is flawed logic, but never mind. The people from Galicia are also known to be quite stubborn and thick-skinned so in Argentina, they use *'Gallega'* for someone stubborn... making this nickname apparently very fitting as it described someone stubborn and from Spain, which they believed I was both. All things considered, I actually accumulated a good few nicknames but I liked them. I'd always been known by a nickname by my close friends and this made me feel like part of the team.

As for the horses, some had perfectly normal names whilst others had funny ones. Some of them had reasons behind their funny or obscure names, whereas others didn't. Normal names included Gerenta, Moncho and Rosita, whilst some of the more obscure names had interesting stories and reasons behind them, like:

Mala Cara – 'Bad Face'. Sounds like a nasty name but simply refers to the broken stripe down her face. She belonged to Cabeza and she was one of the few mares I've ever ridden that I've really loved.

El Colectivo – 'The Bus'. Gillo's incredibly chunky, long horse.

Renuncia – 'Resignation'. When being backed, this horse threw his rider out in the open during a moment of mischief, which scared the rider so much that he resigned on the spot. Ironically, this was perhaps Ruso's only calm horse.

El Último Recurso – 'The Last Resort'. Not a favourite horse by any means, as it appeared. When someone had exhausted all of their horses and had no fresh ones left, they would grab this one, 'as the last resort'. He didn't really belong to anybody, hence the name. He was a stunning animal; a large, strongly built red bay with a wonderful, shiny coat. Sadly, he just wasn't very useful.

El Jubilado – 'The Retiree'. Despite having a reputation for rearing up and throwing himself backwards when saddling him up, when it came to riding this animal he wandered around with his head hanging low like some old man. He was Pirulo's.

Minifalda – 'Miniskirt'. Pavo's mare, which had her tail cut so short at one stage Tío Flaco said it looked like she was wearing a miniskirt.

Pato was the one who seemed to have horses with a bizarre array of names. His horses had the following names:

Vaso Partido – 'Cracked Hoof'. He split his hoof when he was younger and it never properly joined together again. He was the only horse I properly detested because of how painfully slow he was.

Gusano – 'Worm'. He had a short, bouncy step which, combined with the way he arched his neck, Pato said it reminded him of the way a caterpillar moves.

Roseta – These were the horribly sharp grass burrs that were found all over the place. When this horse was being backed, he flung his rider into a dense pile of them.

Come Culo – 'Arse Eater'. Not the most flattering of names at all, but this mare had a reputation for not only being exceptionally fast but for bucking hard, 9 out of 10 times succeeding at throwing her rider off - arse over tit, you might say.

Finally, there was the Alazán Viejo – 'Old Chestnut'. He was simply that, a 20+-year-old horse but you couldn't let his age fool you (more on this later). I inquired Pato because he couldn't have always been called this and when I asked what his name was when he was younger Pato simply replied 'Alazán Nuevo' (new chestnut).

Then, there were others like Por Si Acaso – 'Just In Case', whose name I never found out the story behind, and others who had silly names for no reason, like Enano's very handsome dun number one horse Sapo, which means 'toad'; and Cabeza's stunning little silver dun mare Piojito, which means 'little nit', as in a head louse. Both Sapo and Piojito were *Criollo* horses, which I found very exciting as I had never come across these small, sturdy horses before. To add to the satisfaction, both of them portrayed the pure traditional markings of this breed and had both the dorsal stripe and black stripy legs up to the knees.

At first, using all these strange names was very confusing. Over time, however, it became very normal and actually very useful, considering that El Tío and Pato shared the same name, as did Petaco and Pavo, Gillo and Pirulo, and Cabeza and Gecko's brother-in-law. Everyone had their own unique nickname and this prevented any confusion. Problem solved!

ETCHETO & PROBLEMATIC COWS

Enano, Ruso, Pato, Javier and I had what was to be an action-packed, full on week in the *Etcheto* section, where we had to separate the bulls from the cows and move some cows to another lot. It was the end of February and still very summery so we would be on the horses pretty much at first light, at around 7am, to escape the heat and begin our long rides to where we had to be. As we made our way along, I spent most of this journey chatting to Javier whilst the other three rode closely up ahead. I could feel him glancing over at me and looking in my direction out of the corner of his eye and I could see he was examining my posture, the way I held the reins and the position of my lower leg in the stirrups. Then, as the three boys up ahead kicked into a canter to make up some time, I held Patonegro in his short, uncomfortable canter whilst Javier, who was training up a young chestnut colt he had recently broken in, maintained a fast trot. As a result of the observations we had on one another, we found ourselves discussing the difference between the English and Gaucho ways of riding, in which Javier was totally right in saying that English riding is more about control. On the other hand, the gauchos ride in a more free-flowing, natural way, which is less subtle, in a sense, although ironically they have to be incredibly precise to get their horses to do what they need them to do when working with cattle. One simple mistake or lapse in communication between horse and rider could prolong the job or screw it up entirely.

'We train our horses to work *with* us and over time, a horse that has been well taught will know its job and basically become an extension to its rider as they work in harmony with one another, whereas you train your horses to listen to you and as they progress, you challenge them further, meaning you always keep them thinking,' he remarked rather philosophically. 'It's all about understanding and being in sync with your horse, or else they are no use on a ranch.'

I processed this and he was right. He had stated something which to them is very obvious, but I had never really thought about it before. A good ranch horse knows how to work cattle. That's its job. They're produced for it and do it until the end of their days. As in England, the horses here are backed at about 3 years old and as soon as it understands the basic commands it's taken out to work. Unless they become injured, which out there happened quite rarely, they will work into their late 20's before retiring. Over these years, as they gain age and experience, they improve and even excel at their jobs. In contrast, the average horse owner in England might start backing a horse at 3, but they probably won't start taking it eventing or anything until it's about 5. The belief is that everything must be slow and gradual. Showjumping, cross-country, dressage, maybe hunting - the trouble then is, after all this, you will have either damaged your horse or burnt it out with high mileage and at 15, you will have a semi-arthritic horse to hack out and perhaps, if it hasn't lost its mind, a young rider can do some Pony Club rallies with it. In short, we Brits like our 'all-rounders' and then get upset when a horse that can ace every single aspect is only one-in-a-million, purely because perhaps we don't give our horses time to become experts in a discipline because we are always levelling them up. But, when you *do* manage to get your hands on one of these 'perfect horses', you'll only be able to enjoy a couple of seasons with it anyway before you have to retire it.

'Sofi,' called Javier, to recapture my attention. 'Look, I'll give you an example of what I mean about us having a natural understanding with our horses. See that dune over there? Ride up to the top of it and you'll see three horses over in the distance.'

'How do you know?' I asked inquisitively.

'Because I do, that's the point. You learn to feel them around you and even smell them.'

'You're telling me you can smell three horses miles away in the distance over that dune?' I responded quizzically.

'Stop being stubborn and just go!' he laughed, threatening to give Patonegro a smack in order to get me moving.

'If you're wrong, it's going to be really embarrassing for you,' I teased.

Hesitantly, I veered off the track and trotted up the dune he had just pointed to out to. Patonegro and I made our way up the incline and when we reached the top, three horses that were calmly grazing in the distance lifted their heads, ears forwards, observing us as we appeared over the peak.

I chuckled to myself in surprise and instantly felt my face go red. How did he know? Did he see them when we were further back and, like some sort of magic trick, just so managed to sway our conversation in such a way to achieve this? Were they perhaps some of his own horses that he knew liked to spend time in that particular spot? Or, did he really feel their presence and catch their scent in the breeze...

I spun Patonegro around and we cantered back to the others. Having examined the look on my face, without me needing to say a word Javier knew he had been right and laughed. Despite feeling quite insignificant at that moment and trying to hide under the brim of my hat, there was something about this confident, chatty, smiley chap that made me feel like I would have a good friendship with him.

The first lot of *Etcheto* we were in was beautiful. It was the first time I had been to this part of the ranch and the dunes were absolutely breathtaking: different shades of yellow and green, tinted by the silver sheen of the *olivillo*. This meant the entire landscape shone and glistened. Standing at the top of a dune looking at the infinite undulations through Patonegro's black ears was just so peaceful and relaxing. Right from the start, *Etcheto* was my favourite part of the entire ranch and I always enjoyed working there, just because of the scenery alone. The family had spoken a lot about the beautiful lagoons scattered around the property, which of course I was dying to see, and on this ride I finally managed just that. The first lagoon I saw was the one where the family have their little cabin, where they go to water ski and swim. I got up onto higher ground and there, having been tucked away, it suddenly appeared from between the dunes. It was huge, very open and almost a perfect oval shape. It was about a kilometre and a half long and half a kilometre wide (so a very sizeable body of water), even though I was told it wasn't more than about 2 or 3 metres at its maximum depth. Meanwhile, some of the other lagoons reached depths of almost 10 metres. I thought it was beautiful but the gauchos were saying that however lovely it was, there were others which were far, far more picturesque since instead of sitting like a large lake as this one did, others would meander at the base of the dunes.

As we rode along, a group of rheas appeared ahead of us and almost instinctively, Enano and Ruso darted off after them, followed by Pato, who felt obliged to do the same and then naturally me, who couldn't miss out on this. It quite quickly became clear that this was just one of the things they did for fun. Rheas, being mini ostriches, are incredibly fast and catching up with one on your horse was a challenge all the gauchos tried to achieve. Now, with me there, it was even more fun for them because not only did it give them the chance to show off, but it also gave them yet more opportunities to tease me since I rather foolishly told them stories about my time on the ostrich farm in Australia.

With me being new to everything, I did feel sorry for the horses, as riding in *Etcheto* meant you were constantly riding up and down dunes, which I couldn't help but think must have been incredibly tiring. However, Patonegro and the other horses didn't seem to mind. I soon realised how incredibly fit and good at their jobs these animals were. I loved the relationship the gauchos have with their horses because they are treated as a friend and work companion, not as a pet. This meant they were treated with kindness and respect, but also as animals. Not like back in England, where horses are practically sacred and have an awful lot of fuss made about them. These animals worked hard, but only for a week or a few days at a time, and they would then be set free in a lot to rest and graze in a herd for 3 or 4 weeks. This is why each gaucho had at least 3 horses, so they could rotate. Some horses had their own funny quirks (I soon came to experience some of these) but generally, the moment you had the bridle on the horse it was totally under your command. Sometimes, you would have to dismount and go through some *caldén* thickets on foot to check there were no cattle hiding inside. You'd get off, wander into the trees and when you walked back out your horse would be stood there, exactly where you left it, without even needing to tie it up. You could do this anywhere: whilst opening a gate, in a wide open space, deep within the bushy *monte* or even in the cattle pens, where the horses would happily stand among the cattle. But, the moment you take off the bridle, the horse will leave you to mind its own business. I found it truly fascinating.

Between the five of us we spent about 40 minutes rounding up what must have been four or five hundred cows and calves, which we passed through a collecting pen before moving them on to another lot. This drive was nice and easy for me to start with as everyone was in sight of one another and Pato never ventured too far away, to ensure I was always getting on OK. At the end of the drive, as the gauchos all appeared one by one at the collecting pen, I expressed my joy at the beauty of this part of the ranch.

'Ha,' they scoffed. 'You liked that? Wait until we go into some of the other lots with all the wildlife and the other more idyllic lagoons. Then you'll be speechless.'

We moved into the next lot over to round up a small group of cows. By small I still mean a hundred or so, but compared to the general herd sizes here that was nothing and the trouble with a smaller number of cattle in 544 hectares is that they're harder to find. As soon as we got into the lot, the boys kicked on and all cantered off into the distance and without knowing exactly what I was doing, I improvised and did the same. From the very few drives I had done so far I saw that you herd the cattle by starting at one end of the lot and you ride across it, shouting and making noises to push them along. When they see and hear you, they will move away from you in the same direction you're going and thus, head towards the fence on the opposite side. When they hit the fence, they quite

71

simply walk themselves along it and you guide them so they go through the correct gate. Following this logic, I thought I would do just that and cut directly across this lot, which would position me at the back of the drive, where I would push the cattle along from the back. Sounding like a perfect plan, I kicked Patonegro on and we cantered off, up and down the dunes. The trouble was, I was still very naïve about the size of the land out here and I found myself cantering on and on and on - and no fence appeared. Indeed, there wasn't even a fence in sight. I directed my patient boy (who must have been thinking, 'You have no idea what you're doing, do you?') up the tallest nearby dune and there we stopped, where I allowed him to catch his breath and I looked out across the landscape whilst listening, trying to catch wind of where the others were.

Eventually, I started seeing cows appear over the dunes being followed by horses and I realised perhaps it was for the best that I hadn't continued with my plan because they were all heading towards me, being directed into the handling pens behind me where we started our drive. The gauchos hadn't galloped off across the lot – they took off along the perimeter to get to the other end and bring the cattle back to where we started. If I had gone on with my plan, I would have turned the cows around back the way they came and this would not have made me very popular, I'm sure.

Since I wasn't very far away from the handling pens, I caught sight of Ruso who was galloping ahead in order to get to the pens before the cattle, so he could direct them in as they came and man the gates. This, and the sight of the cattle already heading towards me herded on by the others, sparked an ingenious idea that would prevent me from being humiliated: I would get down from the dune so Ruso wouldn't see me and hide, waiting for these cows to pass me by. Then, I would ride along behind them to make it look like I had brought them along myself. Thrilled with the idea, the plan did work and with a smug look on my face I guided these very few cattle to the pens and in through the gate. Then, to continue appearing busy I turned around and cantered off again to search for any other animals, chuckling away at myself as a result of the success of my plan. A short distance ahead Enano, Javier and Pato were frantically chasing five cows that refused to be rounded up, so I thought it was best to go and help.

There was a small lagoon not far from the handling pens, which is where these cows were heading, so I headed towards there too. As the others tried to herd these cows around the lagoon, I decided to position myself at the bottom of two dunes to block the cows from escaping that way. However, the small herd broke up and a few cows scarpered. Foolishly, I chased after one, which meant another cow ran straight through the path I had been blocking. She then totally legged it in the other direction and I knew instantly that I had messed everything up.

'I'm so sorry!' I said repeatedly as she ran farther and farther away into the distance, feeling truly embarrassed and annoyed at myself.

We managed to bring the 4 cows we had partially kept under control into the pens and then went back yet again to fetch this last rebel cow, which turned out to be far more eventful than it should have been.

She was determined not to be caught. She ran and ran. We galloped and galloped until she started getting tired and, very intelligently, decided to go and stand in one of the lagoons. I had no clue what to do, so I decided from this moment to just stay back and watch. Given that we were in this situation all because of me in the first place, I thought this was probably the wisest thing to do.

Pato and Enano were shouting at this cow, trying to scare her and get her to move out of the water but the clever beast knew she was safe where she was. Javier then tried his bit and was pushing his colt, who was actually working with cattle for the first time, into the water but of course, being young and inexperienced, he had no idea what his rider was asking him to do. With the poor young horse beginning to panic and get stressed, Javier hopped off and decided to try to get the cow out on foot. Currently, the cow was only in the muddy shallows so being the only one in long boots, Javier was fairly confident he could get to her. What happened next was so quick. I can't really describe what happened accurately but all we saw was Javier running deeper into the lagoon. The cow suddenly began chasing *him* and he then disappeared into the water, with the cow passing straight over the top of him. It all happened so quickly and all we saw was Javier's hat floating on the surface of the water. Even the cow looked back over her shoulder to see where he had gone and she looked totally confused at his sudden disappearance, as she stood uncomfortably with her ears pricked and snorting heavily. We all sat rigid on our horses, preparing to jump off and run into the water if he didn't resurface pretty soon, all holding our breath in anticipation, hoping desperately that nothing serious had happened. Luckily, this wasn't for long. With a sudden gasp, Javier emerged from the filthy water like some sort of swamp monster, covered in mud and algae, flailing his way back to dry land. Seeing he was OK, we all creased and wheezed with laughter - which consequently scared the cow, causing her to run out of the water.

Javier's disappearance, the cow's confusion and his little hat just floating on the water were too much to handle. We thanked God that nothing disastrous happened. It's safe to say, poor Javier never lived this down. When we finally managed to compose ourselves, Javier grabbed his sodden hat out of the water and slapped it on his head, both him and the hat still covered in algae. He then trudged out of the smelly water, spitting out bits of weed and mud. He approached his foal, who, upon seeing this dripping, decorated mess approaching him, squelching with every stride, he panicked and refused to let Javier mount up which of course made us all start laughing again. After a few moments he settled the young steed, jumped on and once again we all set off after the cow, who was benefitting from our kerfuffle and thus continued making her getaway. Eventually we caught her with a lasso and given the combination of the heat, how far away she was from where she had to be, plus the fact that Petaco was on his way to pick us up to take us back to the *casco* for lunch, we tied her to a *caldén* in the shade for us to deal with later.

When working far away from the *casco,* during lunch and overnight, the horses are left in smaller paddocks with a water supply, called a *potrero*. Someone would then radio Petaco and he would come in a vehicle to pick us up. We unsaddled the horses, washed them down and let them go. It was a hot day, so by now Javier had dried off and was left caked in a fine layer of dried mud, with the odd green bit still hanging off his hat or a button of his shirt. When the truck showed up, he sensibly hopped into the tray rather than inside.

'Would anyone like some water?' he asked, as he took off his boot and poured about a litre of water out of it.

Even after lunch, this cow did not stop giving us grief. Upon seeing us return, she got aggressive, threatening to charge at us but being tied up, all she did was tangle herself up and wind herself round and round the tree. Despite his previous encounter with this animal, Javier got off his horse and tried to take the rope off but she was pulling against it, making it impossible to remove. Pato then suggested pushing her with the horses so that they could unwind her and untie the rope from the tree, so that's what we tried. Pato's horse wasn't ideal, but I remembered Cabeza telling me how good

Patonegro was at this sort of job so I offered Pato my horse. With Patonegro struggling, Javier then hopped onto Pato's horse to help, since his foal simply didn't have the experience or strength for this. While Ruso and Enano untied the rope from the tree, I kept hold of Javier's foal (who was just as confused as I was). Somehow, eventually, they succeeded in untying the lasso from the tree. Given her behaviour, they decided to walk the cow on a lead to prevent any more shenanigans. Javier rode off, leading the cow with Pato pushing her along. Then, Enano hopped back onto his own horse and followed, leaving me holding Javier's foal. I eyed him up, assuming I would have to get on him, until Ruso approached me.

'Take my horse,' he said to me and handed me the reins to Renuncia. 'Not sure it's a good idea for you to get onto a foal on its first time being ridden out.'

So, there we were. Thanks to this cow's vengeance we had all (apart from Enano), ended up on one another's horses. The interesting thing I found about Renuncia is how sensitive he was to my way of riding. Moncho, Patonegro and Zaino Panzón didn't bat an eyelid when I got on them, but Renuncia must have instantly noticed my different way of riding. He seemed awkward and unsure of what I was asking. I noticed this in a few horses; some were forgiving, or perhaps just didn't care, whilst others really noticed the difference from the moment I got on, which was another thing that really astounded me about these particular horses. Renuncia didn't do anything nasty, but he stood very rigid when I kicked him on and when he finally decided to move he started with a jolt, as if threatening to take off with me. He did eventually relax and once he was comfortable, he remained a good boy and given the story behind his name I was grateful he didn't decide to dump me on the ground and run off. As for Javier's little foal, without any drama whatsoever, Ruso rode him back over to the pens to return to Javier, who felt he had worked quite hard enough on his very first day of work. Not wanting to overdo it, that evening he let him off to have a rest and decided use a fresh horse tomorrow.

The following day we headed to another lot in *Etcheto*, this time to take the bulls away from the cows. As we rode across the lot we were in the previous day, Ruso cantered off by himself. I wondered what he was up to, but then saw him in the distance swapping over and saddling up another horse in one of the *potreros*. When he joined us again I saw it was a young horse; a handsome chestnut gelding with a gentle expression and a star on his forehead, which reminded me a lot of my first horse. As I looked over this horse I noticed that, in place of a bit, he had a piece of soft, thin leather wrapped around his lower jaw This strap, instead of being right up against the pre-molars (where the bit sits) which causes the corner of the horse's mouth to be pulled back, was positioned low down directly behind the incisors, not pulling back on the horse's mouth at all. I had seen this technique in photos but didn't know it was an actual thing that was still done today. I questioned its purpose, to which Ruso explained that it's the common thing to do with young horses; most gauchos will start all their horses with this leather strap instead of a bit on the basis that it's not sharp and it doesn't damage the corner of the mouth since you're not pulling against it, which keeps the mouth soft. Also, with the unpredictable nature all young horses possess, if they bolt or whatever and you need to pull them to a stop, you don't damage their fragile young mouths with a bar of metal. Instead, you pull on their jaw without causing damage since the leather becomes lubricated by the horse's own saliva and it is accommodated in such a way that it doesn't tighten as you pull or anything nasty like that. It also means they don't learn to fight back by taking the bit in their teeth, sticking their head up and running away with you since the strap is situated lower in the horse's mouth, which pulls their chin to their chest and they cannot take hold of the leather in their teeth.

However, that doesn't mean to say it's totally comfortable because it presses down on the horse's tongue if you ever need to pull harder in the event that your horse doesn't respond to gentle pressure from the reins. Therefore, over time, by having its tongue restricted by the leather strap the horse learns by itself to respond to slighter movements, so when they receive pressure from the leather directly to their jaw they learn to reject the harsher commands and give in to the softer ones. This is the sign the gauchos look for, because as soon as the horse starts disagreeing with the leather and reacting to more gentle pressure, they swap over to a bit. Just before doing so, however, (using still the leather strap), they will hold their horses very tightly in the mouth for a while, pulling back and applying unusually strong pressure. Sometimes, they may even lie a horse down and pull their chin to its chest. Then, they look out for a subtle sign of the horse giving in to this and relaxing under the pressure, which it refuses to fight any longer. This may seem harsh but it's only done once and it seems to work. Now, since the horse has learnt how to respond to rein pressure on the neck for turning and soft pressure on the mouth for stopping, the horse has naturally made itself 'soft in the mouth' and as a result, incredibly easy to control. In addition, you have not damaged the horse's delicate young corner of the mouth with a metal bar, but instead have worked on a part of the mouth that will never be bitted or restricted ever again. The result – all the horses were ridden in an Argentine snaffle (usually a solid bar with short shanks) and they carried themselves happily on a long rein which is what made them so lovely to navigate by pressure on the neck alone. I remember once testing just how gentle you can be, that I cantered through the thick *monte* areas holding the reins totally loosely between my index finger and thumb, navigating through the bushes and trees giving such subtle commands it was remarkable the horse even picked up on them.

You see some horses in England ridden in incredibly strong bits and even with this, they fail to respond to you tugging aggressively with both arms. On the other hand, these horses stopped with a gentle pull from one hand – and that's only if they haven't reacted to you altering your position and balance in the saddle. Even the 'stronger' horses on the ranch could be stopped the same way. They weren't bitted with anything harsher and they certainly didn't use additional aids like martingales to stop them lifting their heads, or flash nosebands to strap their mouths shut.

I thanked Ruso for his explanation and saw perfect sense in everything he was saying. After being enlightened to this I started looking at the corner of the horse's mouths when unbridling. It was certain - none of them were hardened or cracked from their bits, whereas back in England there have been plenty of times I've had to put cream on the sore and cracked corner of the mouth of a strong horse with a harsh bit. There was another thing I noticed about this young chestnut that was different from the other horses, and that was his mane. All the horses had their manes entirely hogged, including their forelocks, or 'fringes', as non-horsey people would say. A hogged mane is one that is totally cut back. The only bit of mane that would be left long was right at the base of the neck at the withers (nearest the saddle), so that when mounting up you had something to hold on to. However, Ruso's colt had a section of mane - no more than a hand's width, which was left long - about a quarter of the way down his neck from the ears. I thought perhaps he just liked the look of it, so I didn't ask any questions.

I now knew why Ruso had swapped horses and that was because we were to obtain an extra team member who up until this point I had not met. It was Gillo, Ruso's brother. Having arrived back at the ranch mid-week after returning from his holiday, it was easier for him to use Renuncia than waste time meeting us with one of his own horses. Gillo, in his early 30's, was the elder of the two siblings and in contrast to his tall, slim brother, he was shorter, of a heavier build and had a kind, gentle expression to match his placid, tranquil nature.

Javier did, however, have to bring a fresh horse over from the *casco* and join us at the *potrero* since he had released his foal the night before. When he got to us, I recognised the horse and couldn't believe it. The bay he turned up on was the same one I had seen behaving like a totally wild animal in the cattle pens at the weekend, galloping away from us and bucking hard, before rearing constantly whilst squealing loudly when he was finally caught and had a headcollar slipped on. He was a *redomón* (what they call a horse that's untrained or not yet fully trained) and totally unhandled - when I last saw him, at least. After all the fuss in the cattle pens, when Javier finally managed to catch him, he said he would begin backing him. That very afternoon, he led him into the round pen at the *casco* where quietly, and totally by himself, he got on with training this horse. I joined Gecko and the others to do more work with the calves in the feedlot so unfortunately I didn't get to see him working with the horse but before we went away, I snuck over to catch a glimpse of it and there, with no shouting, whips, or aggressive gestures he was peacefully guiding the *redomón* into his new life as a stock horse. That was on Sunday afternoon and yet there we were, on Tuesday morning, the colt's first time being properly ridden since being trained in a matter of hours on Sunday. He was a totally different horse: calm, completely relaxed and with better than golden behaviour, not only considering it was his first time being ridden outside of the round pen, but also for having been ridden all the way here alone without the company of another horse to comfort him. Cool as anything, he behaved as if he had done cattle work many times before.

'I don't believe it,' I stuttered to Javier as he approached.

'Quietly and with patience, all you've got to know is how to speak to them,' he replied.

I was in awe. Even though I had only known the gauchos for a few days, I already loved them, not only for who they were but also for their connection to nature and their horses. Many like to refer to the cowboys when talking about horsemanship and horse whispering, but definitely, the gauchos do not receive the acknowledgment they deserve. It got me thinking of the day before, when Javier had got me to look for those three horses from the top of the dune... perhaps it wasn't just a coincidence or a lucky guess. Maybe, truly, the connection these men have with their horses and their environment is something totally unique and special to them, something that runs only in their blood and veins and is therefore incomprehensible and inimitable to the rest of us who have not been bred to possess this power.

This next place we worked in was even bigger than where we were the day before, with 1,111 animals in 652 hectares, but I was terribly excited to see this place as I had been told how beautiful it was.

As we passed the gate into Lot 13, Pato pointed to the top of a very steep, tall dune directly in front of us.

'Ride up to the top and have a look,' he said to me.

I kicked Patonegro up the steep incline and on reaching the top, I was utterly speechless by the dunes as far as the eye could see: the colours, the infinite cloudless blue sky and the gorgeous lagoon, all framed by my little horse's contrasting black ears, was exceptional. *'Mi Pampa infinita'* rang in my head as I looked out. I honestly have never seen anywhere so beautiful and unique as this. As I grinned and pulled out my camera I heard a 'told you so' from one of the gauchos down below.

A lot of charging around was involved this morning. Finding just 37 bulls amongst 1,074 cows in such a huge area was not easy. After debriefing and advising Javier to keep well away from the water's edge, we all scattered off in separate directions. If I had felt bad pushing Patonegro up and down the dunes the previous day, this was even worse. Here they were much steeper and much, much taller but yet again, even in the heat, he got on without complaining or hesitating once. At first I shadowed Pato for a while, to ensure I set off doing the right thing and together, we scrambled up a dune. At the top, we looked down at a gathering of a couple of hundred cattle, all clustered together right in the corner of the lot and paying no attention whatsoever to us. They were peacefully going about their business, munching on tufts of grass and drinking from small pools of stagnant water. There were loads of springs all over *Etcheto*, which formed the huge lagoons as well as these little pools that were dotted all around the place. It was obvious that the cattle did not care which water source they drank from.

We stood there for a brief moment without saying a word until Pato broke the silence.

'Sometimes all you need to do is sit and observe. The job isn't all about galloping around as fast as you can at all times. It's pointless. Like that, you miss things and unnecessarily tire out your horses.' He looked over at me, before returning his gaze to the gathering of cattle. 'How many bulls do you see in there?'
I searched hard but couldn't identify a single bull from the cluster, which was condensing now that the cattle had spotted us.

'None at all!' I confessed whilst laughing. 'How many do you see?'

'There are five,' and he proceeded to point them out to me, but even like this I still couldn't see them.

Eventually, he gave up trying and descended the dune to break up the herd to get the bulls out, something I helped him with before setting off on my own. Leaving Pato at the collecting pen to man the gates and make sure none of the animals escaped, I felt confident I could take part in this task and cantered off in search of bulls. In no time at all I found one, but as soon as he knew I was onto him, he ran off the top of a very tall, almost vertical dune and I was determined not to lose him. At the precipice, I gave Patonegro an extra kick and he practically threw himself off the top of the dune. I had to fling my feet forward towards his shoulders to prevent myself from slipping off. As I forced all my weight right back and basically lay there with my legs outstretched to almost either side of his neck, Patonegro also sat right back on his hocks. With no drama whatsoever, he smoothly and carefully made his way down. I must confess, although I was trying my best to ride the same way as the gauchos, the temptation to collect up my reins and keep a good contact on my horse's mouth was very hard to overcome in instances like this. Under any other circumstance, I would never have dreamt of going down such a hill with my reins hanging loose, but being second nature to him Patonegro made his way down easily and once at the bottom, without delay, we continued with our chase. Exhilarated, I felt as if I had just conquered the famous bank at the Hickstead Derby, only this was steeper and much, much taller. After doing this countless times throughout my days in *Etcheto*, I arrogantly started wondering what all the fuss over the bank at Hickstead is all about. We caught up with the bull, got around him, pushed him back up the way we had just come and gathered him successfully into the collecting pen, where Pato shut him in.

'*Sos buena*,' he said with a smile – 'you're good.'

My little black horse amazed me more and more each time we worked together. He turned on a 50 pence piece and galloped fast, never taking his eyes off where he was going or the cow we were chasing. The place was a haven for armadillos, which made perfect hoof-sized holes and tunnels everywhere, and then there were the plains vizcachas – a funny looking rodent the size of a large Jack Russell that loosely resembles a chinchilla. The males are characterised by a rather amusing thick, black, wiry moustache that stretches across their faces from cheek to cheek. These critters would burrow at the base of wood, so under trees or along the fence lines at the foot of the wooden stakes, and produce a cluster of holes similar in appearance and presentation to badger setts, called 'vizcacheras'. They were easier to see than the small holes the armadillos made but innocently, one could think that you could go across them by walking between the holes, without realising that below you all you have is air and that there was only a thin layer of sand between your horse and a deep tunnel down below. Luckily I was warned about all of this, but not all the cattle had learnt the hard way so from time to time, when herding the animals along the perimeter, one would make the mistake of walking across the vizcacheras. All too suddenly, under their own weight the earth would open up and engulf the animal into the network of tunnels below, causing them to struggle and crawl out. Anyway, between the armadillos and the vizcachas, they made this otherwise fairly benign terrain quite perilous. I was always terrified that my horses would put a foot in a hole and tumble, but this was absolutely not an issue with Patonegro, who would hop and navigate around all these holes himself, even at speed. Partridges would fly up straight past his head from under his galloping hooves and they wouldn't even startle him. In fairness, all this was true for just about all the horses I rode out there and it would make me think of how truly pathetic my own horse was. Once, he nearly hurled me into a show jump because the wind blew which caused the daffodils around the arena to sway in the breeze, which frightened him terribly. As for a pheasant running out from under a hedge, this was a guaranteed way of putting a bitter twist into any hack in the countryside.

As I was riding away from the pens after depositing the bull I had been chasing, six cows and another bull ran up over the summit of a dune. Upon seeing me, they stopped on the flat ground not too far from the pen itself. They stood watching me grouped together, the cows surrounding the bull. I glared at the bull and he glared at me, and one thought went through my mind. I've seen many videos of it, as well as at rodeos in the States, but in practice I had no idea how to do it... but I was sure as damn it going to try; *I'm going to cut the bull out.*

I felt that if I was going to earn the respect and a place on this ranch amongst the gauchos, I needed a strong start and to appear useful, so I gently walked Patonegro towards the cattle, causing them to shift. Appropriately timed, Ruso, whom I presumed had been the one to push these cattle in this direction, appeared over the brow of the dune exactly where the cattle had come from. He stopped dead in his tracks and watched, anticipating what I was going to try and do. Knowing I only had one shot at this, with only a matter of seconds to do it in, everything I needed to do had to be spot on so I ignored him, pretending he wasn't there so I could focus. I began riding perpendicular to the cattle and the more I approached, the more eager they became to slip away and run. As they fanned out preparing to run, I jilted Patonegro towards them, aiming for the bull who was in the middle of the line. This encouraged the four cows at the front to charge off and run away, whilst my positioning blocked off the bull and two cows from following them. Knowing they had become my targets, they began to get shifty. Desperately wanting to follow after the cows who had just managed to escape, the two I had under control walked past the bull, leaving him at the back and positioning themselves perfectly for me to cut him out. I stood very slightly ahead of them – not enough to be in front, but sufficiently so to place pressure on them and hold them in place. This is occasionally referred to in farming as the 'livestock force field', which is just knowing where and how far from the animals you need to place yourself in order to provide enough pressure with your presence alone in order to hold them still. As a general rule, the more feral and poorly handled the animal, the further back you must stand to hold them in place without causing them to panic and run off in another direction. In this case, working with semi-wild cattle whilst on a horse, I was a fair distance away from them - some 20 metres or more.

With my force field applied, the three once again stood solid and unsteady, staring at me with angry, anxious eyes and their big ears forwards, grunting as they flicked their tails aggressively. This next bit was my final move. As a result of these cattle being so wild, which caused me to activate my force field from quite so far away, I was going to have to be very quick to cover the ground between us if I was to succeed, else the three of them would take advantage of the distance and scarper. With Ruso still watching quietly and inquisitively, my determination was strong. Patonegro had his ears pricked the whole time and hadn't taken his eyes off the cattle for a second. This meant that when I gave him a firm kick, he launched himself forward, hurtling at high speed towards the cattle. Naturally, the three animals reacted by running but with my horse's quick reactions and speed, we caught them off guard and were able to get between the cows and the bull, forcing him to stop dead in his tracks. As the two cows ran away to my left, I successfully cut out the bull and forced him off to the right, where I then herded him straight into the pens. I was pretty exasperated, given that all this happened in a matter of seconds, but I was gleaming with success. As this bull joined the others inside the pens, from over to my left, I heard a simple *'wep'*. As I turned to look, Ruso gave me a gentle nod, spun his colt around and galloped off down the dune out of sight. 'That must have been it,' I said to myself. I had just received my Gaucho Seal of Approval.

Having been used to working with cattle in England, I always perceived these animals to be quiet and gentle, with the occasional feisty or angry one. There, it was the complete opposite. Treat them *all* with caution and every once in a while you'll come across a quiet one. Unlike cattle in the UK, those on the ranch lived wildly and aren't handled often, so they're not used to human interaction. After some time, all the bulls had been gathered together and successfully penned up – all apart from one. The boys were surrounding an angry black bull from a distance. As I observed him more closely, I noticed a large lump on the top of his head which had granted him the name 'The Martian'. It was clear from this that they were familiar with this animal, and that they were. When I asked about this strange lump they said it was a type of abscess he'd given himself from head-butting everything he didn't like, which appeared to be most things.

'*Guarda, Sofi.* This one is really bad. He'll charge at the horses,' Enano warned as I approached.

That he most certainly did. As soon as I got nearer, the imposing threat of an extra horse challenged him and he charged at us, instantly forcing me to spin around and gallop away until he stopped chasing me. After a lot of charges at all of us and a chorus of 'Wey hey!' that accompanied each time he did, all the bulls were eventually penned up in a small enclosure and the horses were left to rest at the windmill whilst we headed to the *casco* for lunch. I had been told to get some rest as we had a long ride in the afternoon to herd the bulls to their new pasture.

They were certainly right about the long ride. After a lot of searching and help from the gauchos at lunch, I managed to find the ranch on an online map so I could get an aerial view of it. I was interested in measuring the distance from the *casco* to Lot 13, as it is right in the far corner of *Etcheto* against the neighbouring property, *La Gitana*. Also, it is one of the furthest points away from the *casco* on the entire ranch. The result was that the *casco* was 16 kilometres (10 miles) away as the crow flies from the *potrero* we had left the horses in. Hence, riding this distance would have more than doubled seeing as the tracks most certainly did not cross the ranch in straight lines.

Due to the strong afternoon sun (it was about 34°C), both for the sake of the horses and the bulls we went along slowly, so as not to overtire any of the animals. Although the temperature was high, it was completely dry and there was no humidity whatsoever, despite the lagoons dotted about. Thankfully, this made the heat very tolerable, to the point that it wasn't all that noticeable. Perhaps having arrived to Argentina directly from Australia also had a part to play, giving me the upper hand in tolerating the heat. Unlike in the Outback where the heat was insufferable and the sun actually stung like needles in your skin, the sun in *La Pampa* was kinder and combined with the fresh breeze I found it very pleasant. It also wasn't extremely hot all day either; the temperatures peaked at about 1 o'clock and by that time we would all be out of the sun having lunch and then of course escape the heat with a *siesta* before heading out again at around 4. At this point, it was already much cooler and the late evenings, as well as the early morning, would bring a slight chill to the air. It was the total opposite of the Outback, where one morning we started drafting cattle in the pens at 4:30 in the morning and the temperature was a fresh, invigorating 24°C. Come 5:30, I watched the sun come up and as it rose into the sky, we could feel the air becoming hotter and more suffocating by the minute. When we finished loading the cattle onto the trucks and went in for breakfast at 8 o'clock, at this point it was already 36°C.

We crossed field after field, with someone riding ahead from time to time to open up a gate. Despite keeping up a good pace, we weren't going to make it to where the bulls had to go before it got dark, seeing as the sun was dropping from the sky quickly. Watching the sun go down in this part of the ranch was simply spectacular; the sky changed colour from the bright blue colours of day to the evening colours of pink, yellow and blue layered across the horizon. As the sun sank deeper below the dunes, it left them silhouetted against a bright scarlet sky until Venus became bright and clear. From here on, the darkness quickly settled. It was amazing how the air suddenly cooled as soon as the sun hid behind the dunes, changing the temperature from an embracing warmth to a pleasant chill. It was enough to make you feel the cool air, but comfortable enough to not need an extra layer. Overnight the bulls and horses were kept together in a sort of *potrero* with a lagoon to drink from, except this one was quite a bit larger than the usual *potreros* – it was more like a collection field for stock than somewhere to leave the horses. We unsaddled, leaving our *recados* on the wire fence and we hopped onto our horses, bareback, down to the lagoon to shower them off. By now, there was only enough light to see outlines and minor details and the horses drinking from the lagoon, silhouetted against the water, looked like a painting. By the time we had all finished, Petaco arrived with the truck arrived to pick us up at 9:15pm. What a day it was and what a phenomenal introduction to the work I would be doing there. Surrounded by such beautiful wilderness, those long days could hardly be considered work – they were more like a gift.

We spent the rest of the week working in *Etcheto* and I enjoyed every moment, particularly how relaxed and sociable the work was. When rounding up the cattle it was just you and your horse, but when getting from place to place with the long distances in between, it gave you a chance to have a good chat to everyone, usually just talking about silly things and laughing. One afternoon, Ruso and I were left to push some calves from the back and I decided to try and break the silence. As a fanatic of music, I decided to use this as my conversation starter. Over the years, I've found this is a very good way to spark conversation. If they love music, you can have a long conversation about everything. If they don't like music, then you know right away that exchanging any words with this person isn't worth the effort.

'What sort of music do you have here? What do you listen to?' I started off.

'Anything,' he replied, 'but mostly *cuarteto* which is the style of music originating from where I come from, in *Córdoba*, which is the province north of *La Pampa*. *Cumbia* is from there, too.'

'Oh, I don't know of it,' I replied. 'I suppose I'm a fan of the 80's. Do you know much English music?'

'I know a few songs,' he replied, bluntly. This was followed by a period of silence until he broke it with a shrill, high pitched 'AAAAA-EEEE-AAAAAA', so universally recognisable that despite it being so piercing and severely ear-destroyingly out of tune, I could still identify it as the chorus of the famous song by none other than the one and only Whitney.

This explosion of shrieking, along with my outburst of laughter caused Pato to find us and ask what on earth was going on.

'We're talking about music,' I told him.

'Yes, yes, that definitely sounded like music...' he replied sarcastically.

'What are you talking about? That was a top class performance I just gave!' scoffed Ruso.

'Do you know or like any English songs?' I asked Pato.

'Umm.... yes, a few actually. Queen, Bon Yovi (the Spanish language doesn't pronounce the letter 'j' as we do), ABBA – ' I was loving this; he was basically listing my favourites.

'*Ero Smí.*'

'What?' I replied, totally confused.

'*Ero Smí,*' he said again.

'Can't say I know who you mean... can you sing or hum one of their songs, perhaps?'

'Erm... no. Probably not,' he chuckled, also blushing slightly, which suggested to me that even if he did know the words, he wasn't going to sing for me.

I sat there, wondering. I tried to think of band names and said them to myself in a Spanish accent to see if any of them matched. Eventually, I got it.

'Ah! Aerosmith!' I rejoiced.

'Maybe, that sounds like it could be it,' said Pato.

'Only 'maybe'?' I replied, to which he nodded.
I'm sure it was probably part of a devious plan. 'Can you sing us a bit, so I can know for sure?' he grinned.
I stuttered, but I was left with no choice so I burst out: '*I don't want to close my eyes, I don't want to fall asleep –* '
Ruso started laughing.

'*Si! Si!* Them! I like them,' Pato responded with a massive cheesy grin across his face and the two of them looked at one another and chuckled.

It was a week of long days and hard work but by the end of it, we rode back to the *casco*. The tired horses quickened their pace on the way home, knowing their shift at work was done. Riding into the *casco* after working away for days at a time felt to me like knights re-entering the castle after completing their quests. You would ride in proudly as the others would watch you arrive from the *matera,* where they would be slurping *mate* or already passing around a beer. It was Friday again, so like every weekend, only a select few would be staying at the ranch whilst everyone else went into town. The family had also left for the city since Lucas had to return to school, so it was just me and the gauchos for the weekend. Juli very kindly gave up the opportunity to go home so she could keep me company and make sure I was well looked after over the weekend, which I thought was incredibly sweet. It was these little actions from the people there that made me feel so happy and relaxed with them. They really did make sure I was well looked after and it was a very genuine sentiment, as they did it from the goodness of their hearts.

After saying goodbyes and wishing everyone a good weekend, I went off to have a shower and prepare myself for dinner. It was while sitting in my room afterwards that I heard a group of voices in the kitchen, so I popped next door to see who was there and what was going on.

I walked in to see Juli preparing a pizza, but the surprising part was seeing everyone else around her: *empanadas* were on the menu as well that night and everyone had gone to the kitchen to help Juli prepare them. *Empanadas* are a typical food in Argentina and are similar to a Cornish pasty, filled with minced meat and then fried. Sometimes they would be baked in the oven, but this was less common. Enano was rolling out and cutting the pastry, Pirulo was adding the filling and folding the pastry over and then passing them on to Gillo, who stood at the cooker doing the frying. This, to me, was such a lovely sight. At the station in Australia, dinner was at 7pm. We would have to be clean and showered before eating the food prepared by my boss's wife and Chanelle. At the ostrich farm, I had to feed myself. Therefore, walking in and seeing this – all the boys helping Juli out simply because they wanted to and enjoyed doing it, brought a big smile to my face. They all loved Juli and she was good to them too, even though they annoyed and teased her at every opportunity. In response, she would barge them out of the way or threaten to hit them with a wooden spoon or something. I decided to get stuck in, so I gave folding the pastries a try. Needless to say Pirulo was a bit of a pro at this, whereas I was not at all. They kept opening and one almost released its contents whilst Gillo was frying it, so Pirulo and I swapped places and I added the filling, leaving him to do the folding.

The evening was warm and the five of us sat down to have dinner together at the little stone table in the courtyard outside the kitchen. With a couple of beers between us (similarly to the *mate*, it's a matter of 'pass the can'), we ate, chatted and laughed well into the darkness. The perfect end to a perfect week.

LAS NUTRIAS

Meal times at the ranch were always enjoyable. We would sit and wait for Juli to call us on the radio and one or two people would go and collect the food. Everyone would serve themselves and take a seat wherever they liked, usually with those they got on best with. There were two tables; one inside the *matera* and one on the porch, where generally everyone chatted and continued teasing one another. When it came to constantly winding one another up, mealtimes were no exception. I would try to swap between eating inside and outside so I would get the chance to sit and chat to everybody. Also, by doing so it wouldn't appear like I was taking preference over who I would sit with. However, I found myself more often than not sitting out on the porch so I could enjoy the fresh air and watch people come in and out of the *casco*, even in the cool evenings at dinnertime. It was easier to speak outside too, seeing as those inside often ate with the TV on, usually watching a programme called *'Jineteando'*, which televised *jineteadas*, or whatever film would be on. One weekend, the small group of us sat having lunch together indoors to escape the unbearable flies and on TV they were showing the film *'Titanic'*. I shared my unpopular opinion about the film, remarking that the ending is silly and unnecessary, given that there was more than enough room on that door for them both to have stayed out of the water until they could be rescued. Pirulo asked me if I had ever cried when watching the film, to which I replied 'no', which resulted in him telling me that I am a cold, heartless human being. As he spoke, his eyes began tearing up at the thought of Rose letting Jack sink into the freezing water.

Meals were a chance to have a proper chat to everyone. We talked about anything and everything, and they all spoke to me in turns and asked me questions while I did the same.

'So, how are you enjoying everything so far?' asked Brandizi.

'I love it. The work is amazing and it's just so beautiful here. You're blessed to live and work somewhere so special,' I responded with absolute sincerity. 'Who knows, perhaps I'll get myself a ranch in Argentina someday.'

As I spoke the last sentence, a sudden blanket of silence fell across the table. Some of them even froze in mid-action, about to stick a fork in their mouth or swallow their drink. I looked around and seeing everyone suddenly quiet and still, I became very confused.

'I'll come and work for you!'

'And me!'

'Me too!'

'To get residency in Argentina you have to marry a resident... You can marry me, if you like. You know, just so you can live here, if you like. You go about your business, I'll go about mine - but at least you'll be able to live here,' remarked Pirulo, trying to be cool and nonchalant. 'As long as I can work on the ranch as well though,' he added. 'That way, as the owner's husband, I'll be able to boss all you lot about!' he cackled as he pointed to everyone else at the table.

'No! That's not fair!' they all started, obviously angry that Pirulo got there first.

'Well, she may not want to marry *you,* in which case, she can marry me,' added Enano with a massive, cheeky grin.

'Or me!'

'Or me!'

And so it went on, all arguing with one another as to why I should choose to marry them rather than anyone else. Whilst this took place right before my eyes, all I could do was sit back and laugh as everything unfurled. Eventually I got them all settled down and said that, even if I did get my own ranch, however much I might want to, I couldn't employ them all because I was sure Miguel wouldn't be too pleased if all his staff packed up and left. With this, they all huffed, groaned, muttered away and carried on eating whilst I continued chuckling away at myself over what had just happened.

The following week, the entire cavalry was needed over at *Las Nutrias* (which translates to 'The Otters') to help Gecko shuffle some cattle around ready for the vet who was coming the following week to do pregnancy testing. Pregnancy testing, or PD testing as it's often called back in England, was done very differently over there compared to what I have experienced in England. Back home, some of the vets used a small endoscope-type camera with a little light on the end which they held in their hand as they stuck it in the cow. They would then be able to see the image on the lenses, which were like small screens, of a special set of glasses the vet wears as they do this. This way, they could see the exact stage of pregnancy and tell you how many weeks in calf the cow was, or in how many weeks she should be expected to calve. They could even go as far as seeing whether there was some sort of problem or malformation. In Argentina, however, the vet didn't have this luxury and on a car ride over to the pens together one day, I asked what it is he feels for. He told me that, with years of experience, your readings become more accurate and that all his predictions are based entirely on what he feels, whether that's the calf itself or the cyst-type growths, called cotyledons, which appear on the placenta and change in formation and quantity at different stages of pregnancy. Depending on the stage of pregnancy, the cows were then sorted into '*cabeza*' – head, '*cuerpo*' – body or '*cola*' – tail, and obviously, any that were barren (not in calf).

Cows, like humans, have near enough to a 9-month gestation period and are allowed to run with the bulls for 90 days. Therefore, the *cabeza*, *cuerpo* and *cola* relates to the stage of pregnancy. *Cabeza* were late stage pregnancy, so cows that got into calf in the first month of being with the bull, whereas the *cola* were the cows that were served at the end, in the third month. As the vet shouted out each stage of pregnancy, the person sitting on top of the crush holding the cow's tail would trim the tuft of hair specifically to match, as an identification method in case the cows got mixed up. An untrimmed, partially trimmed and totally trimmed tail each meant a different thing. In England, you'd have a little notebook and next to the cow's eartag number you would write everything down, whereas on the ranch, the cattle didn't routinely wear eartags because living wildly would mean they would be highly likely to lose them and rip them out. Some were tagged, but not all of them. Therefore, the tail trimming was vital and a simple tally would be kept in order to count the number of barren cows, *cabezas*, *cuerpos* and *colas*.

Las Nutrias was about 12 kilometres by car along the tracks from the *casco*, so riding there was a similar distance, despite taking a few shortcuts across the fields. Nonetheless, it was about a forty-five minute ride over. Enano, Gillo and I took our horses down on

Sunday, whereas the others rode over on Monday morning upon arriving from town. There were nine of us in total: Gecko and his brother-in-law, Gillo, Enano, Pato, Ruso, Pavo and myself. There was also a new member whom I hadn't yet met, and this was El Tío.

Las Nutrias was a funny place as it was incredibly varied over a 'small' area. It had a combination of both flat, open land and some of the thickest scrub areas on the whole ranch. It was as a result of these thick areas of vegetation that so much horsepower was required to seek out and move the cattle. A few men alone would never have managed to find all the beasts amongst it, especially when the cows would amazingly push their way through the thickest and thorniest patches to hide from us. The horses were too big to get through and we would have been ripped to shreds trying, yet the cows seemed to be able to walk through the branches covered in two-inch thorns without so much as a small scratch. This was why the extra numbers were required – to ensure we could steer the cattle away from these dense patches.

For this week I had been assigned Mala Cara, another one of Cabeza's horses. She was a chunky bright bay mare, about 15 hands high and named after her defining feature: the broken white stripe that ran down her face, from her forehead, between her nostrils, to the tip of her upper lip. Despite never being particularly fond of mares, she was superb. Like a lot of mares, she had a strong 'can-do' attitude but was notably kind. She was fast and powerful, but not used to the rider's disadvantage; she had the sort of power that came up from her well-muscled shoulders, which bizarrely would almost send you off the back of your *recado* when cantering up a dune due to the force with which she travelled forwards, thrusting her forelegs forwards with huge power. Anyway, she was perfect in the thickets as she navigated her way through anywhere. Even when pushing our way through thick areas of *caldenes* she would never moan. Even when I hung off of one side of her, one leg over my *recado* and the other under her tummy as I firmly hugged her neck so as to avoid the low thorny canopy, she would simply keep walking in the direction she was pointed in, swerving between trunks and following the path she alone picked out to have branches high enough for her to walk under.

The cavalry rode along, laughing and chatting together, even mucking around. At one point Ruso spun around, put his arm around his brother's chest and pulled him off his horse, gripping him tightly as Gillo hung down the side of Ruso's horse. Meanwhile, Gillo's riderless mount casually followed behind, almost oblivious to the fact that he no longer had a rider. After a few seconds, Ruso returned his brother to his horse and carried on as if nothing had happened. It was all just done for a laugh. The horses were all so placid and desensitized to everything; it was quite refreshing to be able to muck around without worrying about scaring them. Do that sort of thing in England whilst out hacking and you'd probably get a visit from the Air Ambulance and a search party out looking for a hysterical riderless horse.

During our first roundup I stayed near Gecko, or rather, he stayed near me to ensure I was okay, knew what I was doing, and that I was heading in the right direction. At one stage, we rode along together just having a chat and I noticed his horse had the same long tuft of mane in the exact same place along his neck as Ruso's chestnut colt did. Ruso and Gecko were good friends, but this seemed too coincidental to be that they both liked the detail. So, this time, I asked about it. Gecko explained to me that this was a marker to identify a *redomón*. Until a horse was completely backed, it bore this long patch of mane in order to signal that it wasn't yet fully trained and therefore warned other riders more than anything that there was a young, inexperienced horse in the crowd.

'I might leave it on him though, I like the look of it,' he added as he leant forwards and stroked his horse's strong, smooth neck.

He was stunning: a dark bay, only 3 or 4 years old, with a beautifully kind, gentle nature. He had two symmetrical white socks half way up to his hocks on his hind legs and a small white star on his forehead, followed by a small snip on his nose. He was built like a thoroughbred and his muscular body was covered in a silky-smooth, unblemished coat. Gecko had only recently backed this horse, so he frequently used him in order to get him accustomed to working with the stock. He was making amazing progress and Gecko worked this horse as if he really knew his job already.

Gecko and his family were very good hosts whenever we helped him over at *Las Nutrias*. The young couple, their one-year-old daughter and Gecko's brother-in-law lived at the *puesto* together. When we returned from working the cattle, there was always *mate* and warm, freshly cooked *tortas fritas* waiting for us. What's more, the *puesto* had a plant growing there that wasn't found anywhere else on the property, a plant they called *burro*. Gecko's wife added *burro* leaves to the *mate* and it gave it a delicious taste of spearmint. We gladly waited for Petaco or Cabeza to come and pick us up. It is safe to say, with so many of us out working together at one time, the car journeys did become a bit crowded. Having six of us meant there would be one in the front passenger seat and depending on how many people were willing to go in the tail of the pickup, either 3 or 4 of us would be squashed together in the back seats. Foolishly, the first time I got into the truck I placed myself in the centre seat so got inevitably very squished... but the good thing was that I never had to get out to open any gates as a result. This realisation triggered passive laziness in me, making me opt for the centre seat every time we went in the truck but the boys soon became wise to my ways and one day I was forced to the outer seat. When I got out to open the gate, they were so overwhelmed with joy that a photograph was taken of me opening the gate, which was shared with everyone at the *matera* that lunchtime to prove I had undertaken such a task. I would quite happily go in the tray of the truck a lot of the time but in the cooler mornings and evenings, it was always much nicer to be squashed in the back central seat, sapping the warmth from those on either side of me and burying myself behind overlapping shoulders.

Miguel and the rest of his family had returned to the ranch at the end of that week from *Buenos Aires* to continue business at *San Eduardo*, in preparation for the imminent start of the shooting season. In addition, it was also because they had a small film crew coming to record a few clips and take photographs in order to create some marketing videos. I was very excited, for I had been asked to take part in one of these clips.

Mala Cara was brought up for me from *Las Nutrias* and early the next morning, Ruso, Javier and I were to take the horses to one of the lagoons where we would then be met by the film crew, Miguel, Patricia and one of the early hotel guests who had come for the hunting. The idea was to take pictures at the exact moment of sunrise, the two gauchos and I left the *casco,* leading the horses for the rest of the party in total darkness. I had never ridden in total darkness before, so this was yet another new thing that was quite exciting for me. At first, I was quite aware whilst riding along in case my horse tripped up on something I couldn't see, like some stray fencing wire or armadillo holes, but this was silly of me because horses have excellent night vision. The stars were still visible and thanks to the absolute lack of light pollution, my eyes quickly adapted to the darkness and we rode onwards, following the light sandy track to the lagoon. By the time we arrived at the lagoon, the darkness was breaking and shortly after we arrived one of the trucks turned up with the *recados* for the horses we had brought along. As we tacked up, it wasn't long before a second truck arrived, this time with the other film stars aboard. With the sun on the brink of coming up over the horizon, we promptly mounted and at the cameraman's orders, got into position.

The morning was amazing; at first, shots were taken of us all riding in single file up a dune with the horses and riders totally silhouetted against the bright orange backdrop of the morning sky. Then, once the sun had come up fully, we rode through the fluffy pampas grass that grew in great big clusters around the lagoon and with a drone buzzing around overhead, we also rode through the crystal clear water - because it would have been foolish to miss that. Afterwards, we sat at the hut on the water's edge, drinking *mate* and eating sweet treats. This was all filmed as well, of course. It would be months before we would get to see the final video, put together and set with music, but it was a beautiful start to the day and I had no doubt that the wait was going to be worth it. This landscape was beautiful as it was, but seeing it captured artistically through the lenses of professional photographers and drones brought out something even more special.

That afternoon we had to go back to finish work in *Las Nutrias*. The group had split up since those not filming went on to move some bulls elsewhere. This meant Ruso and I would join Pavo, Gillo, Pato and El Tío in helping Gecko. I was told to leave Mala Cara at the homestead as Gecko had a horse down at his place which he could lend me. I chucked my *recado* in the back of the truck and we headed on down. It was hot, just over 30°C after lunch and there were warnings of thunderstorms, which seemed rather speculative when I looked up to the clear, cloudless sky. When we arrived down at *Las Nutrias*, I asked Gecko which horse I could borrow.

'That big grey mare, over there,' he pointed. 'Watch out though, she's totally crazy,' he laughed.

I knew he was joking, for his brother-in-law had ridden her the day before and she was soft as butter. Aside from that, these chaps never gave me anything less than a good, docile horse.

'What's her name?' I asked. I always liked knowing the animals' names and I asked it each time I was given a new mount.

'She doesn't really have one... We just call her *Tordilla* (grey mare).'

A cheeky grin then swept across his face and he looked at me out of the corner of his eye.

'We shall call her Sophia!' he exclaimed. I never managed to come up with a name suggestion that would change his mind so that was that. The *tordilla* was now Sophia... I was Sophia riding Sophia.

As we were out working, the sky began to change. The clouds rolled in and the sky got darker. One could argue either that we timed it perfectly or that we were simply very lucky. At the exact moment we closed the gate behind the cattle, an impressive fork of electric blue lightning came down from the sky, accompanied by an almighty roll of thunder. I felt the tremble resonate through my chest, yet once again, the horses didn't even stir. The air grew heavy, hot and humid as the storm drew nearer, even though it looked like the rain wasn't going to hit us for a while yet. Regardless, no one liked the idea of unsaddling in the rain so together, we cantered back in the direction of home. But, with these gauchos being absolute nutters, the canter quite rapidly turned into a race and when Gecko and Ruso broke out into a mad gallop and disappeared into a trail of dust and, well, I didn't want to get wet either, did I?

With very little encouragement, my fat mare took off. The horses were trained not to follow those in front so even if someone galloped away from right in front of you, you still had to ask your horse to do the same. With a gentle kick we bolted off too and with heavy hooves pounding at the ground we caught up fairly soon and followed behind. When it comes to horses, you can't let looks deceive you. She may have been large, heavy and a bit fat, but Sophia could run. It was actually during this silly race that I got my first proper look at El Tío. Despite being introduced to one another and working together the day before, I hadn't really had the time to get to know him but as we were galloping, a horse suddenly appeared beside me. As I looked over at the rider he was just sat there, perfectly upright with the look of a naughty schoolboy on his face, grinning proudly at me. Despite his horse galloping forwards fully stretched out, El Tío sat there completely still, perfectly in tune with his ride to the point that I could have balanced a *mate* on his head and it wouldn't have fallen off. After a few seconds of grinning at me without a word, he then shot off, leaving me in the dust. El Tío was in his early 50's, with several broken bones and metal plates in his body from various horse and cattle-related incidents yet he was an immensely enthusiastic worker who was always ready to go and full of energy. He always looked fresh as a daisy, was always one of the first to get up in the morning and after the *siesta*, and seldom complained about any aches or pains, which even then did not inhibit his ability or desire to work.

We narrowly escaped the rain which reached us as we drove back to the *casco*, where we on the inside laughed at Pavo and El Tío who had, by choice, decided to go in the back of the truck instead of finding a way to squeeze in with us. We would turn around and tease as we looked at them through the back window, both squatting as close to the window as they could to avoid the needle-like raindrops stinging them in the face. As you can probably imagine, being out in the middle of nowhere meant electricity and Wi-Fi were rather frail and sometimes, the slightest weather pattern could cause it all to collapse. Despite it being only a minor storm that simply passed by, we lost everything overnight but thankfully, the lights were back on by the next morning. However, to make-do that night, in the *matera*, a light was connected to a car battery whilst in the house and kitchen, the food was prepared and eaten using torchlight, with poor Juli in her kitchen with a couple of lanterns and a head torch. These blackouts (there were several whilst I was there) were somewhat refreshing. Limited lights and no internet meant that the *matera* became even more sociable. After a cold shower (because the hot water system was also down) and washing off all the sand I'd had thrown up into my face, hair and neck during our race I walked over to the *matera*, where I heard an awful lot of shouting and laughter coming from that direction as soon as I opened my bedroom door. When I got there, I saw six of the gauchos sat around a table playing a game of cards. I watched this funny game with confusion and asked what it was.

'It's called *Truco*,' replied Chaque. 'A funny game, played with a Spanish deck of cards rather than the standard one. It involves a certain element of lying, misleading and an awful lot of shouting.'

I watched and watched, trying to get my head around the game and its rules, but I simply couldn't. As I understood nothing, there's not a lot I can say about the lying and the misleading of your opponent, but the fact about the shouting was absolutely true.

'I'd like to learn. How do you play?'

'It's complicated,' answered Chango. 'You're better off to watch and learn before having a go.'

I tried and tried but after seeing kings killed by twos and sevens losing to threes, I gave up. That, combined with the diverse lingo, was too much to keep up with. Therefore, I diverted my attention from the game itself and instead eyed up the players, as I was determined to select a tutor and play this game with them. Cabeza wasn't actually a bad teacher at all, but he was cunning and so good at the game that he genuinely knew which cards you had in your hand. He started off as my teacher, and a good one too, but eventually I got frustrated of him saying 'Now play your king' or 'Whatever you do, don't put down your 3', despite there being no way whatsoever that he could have seen what cards I had in my hand. I even checked the backs, in case there were defects or marks on them that would give them away, but they were totally unblemished. Often he'd use his power as a way to help me, but knowing what cards I had also meant that he knew exactly how to destroy me, which was much more fun for him. To add to this, he also got easily frustrated and was fairly competitive, so in the long run, he didn't work out.

Ruso wasn't the most patient type and, being a joker and always one to tease people, I assumed he would be the most likely to mislead his opponent. A good player, but he wouldn't do either. Pato would get excited and his voice would come close to reaching a pitch only dogs could hear which, along with speaking faster and faster, meant I couldn't understand a thing. Plus, we always found ways of winding one another up, which meant we would probably get distracted all the time. I needed someone patient who would be good at explaining as we went along. In the end, down to temperament and ability, I narrowed it down to Chango or Pirulo. One of them would become my tutor in learning this bizarre game.

I loved this aspect of life on the ranch. Yes, you were working at the crack of dawn (sometimes preparing the horses in the dark to set off at first light or before), the days were long and you worked hard, but having a good rest after lunch made all the difference. When the summer days were long and hot, lunch would be at 12 and you'd start work again at 4. Then you could be out until 8 or 9pm - back for dinner - and people would hang out at the *matera* for a while afterwards, either just chatting or playing games and watching television. There was no rush to go to sleep. People would clock off and head to their rooms at around 11pm. It made the whole atmosphere very sociable, which is what I liked so much about it. In Australia, at the cattle station, we had breakfast at 6am, stopped half an hour for morning tea ('smoko'), then lunch and the day would then finish at around 5:30pm, having worked through the extreme heat of the day. At 7 o'clock, dinner would be ready and by 8, we would have eaten, headed to our rooms and fallen asleep by 9:30. At the ranch, however, people had the energy to stay up and spend time together and I loved it. It gave the place a very warm feeling and was probably the reason why everyone there got on so well. The positive attitude kept spirits high, which I'm sure had a direct impact on the enthusiasm they had for their work.

GUNSHOTS & ENGLISH LESSONS

The shooting season was, as much for the family as for the gauchos, the most highly anticipated time of year. With the first guests of the shooting season arriving imminently, I put myself to good use and gave the gauchos a few basic English lessons. A group of hunters from the States were due to arrive to start off the hunting and I had already spoken with the family and gladly agreed to be tentative to the radio at all times, should anyone need any communication or translation help.

As to be expected, some of the men were much more enthusiastic about these lessons than others. The weekend before, I had spent a few hours going over basic day-to-day and hunting-related vocabulary with Pirulo and Enano, which went rather well. After coming back from work, I sat with them outside their room sharing some *mate* when Pirulo asked me a few questions, like 'How would you say this in English?', so it seemed like a good time to tutor them. Enano was definitely the shy student, but he paid close attention and seemed to take everything in. Pirulo, on the other hand, was very engaged and not afraid of speaking and pronouncing.

Since it was him who requested the lesson, he came out with a notebook and wrote down all the translations I put before him. He actually did well from these lessons and walked away with a couple of pages' worth of notes, which I was very proud of him for. Pato and Petaco were also pretty good, even though Petaco felt he was a pro and had the upper hand simply because he was fluent in the word 'beer', despite calling the black-bucks 'blah blahs'. Trying to teach all the others, however, was a bit trickier. There were those that had been hunt guides for several years and had somehow managed to get by, so they didn't feel the need to start learning now. Others felt they knew enough based on what they had picked up from previous years. The majority of the gauchos had registered to be hunt guides, leaving only Chango to his groundskeeping duties, Brandizi at the hotel and Pavo, Gillo and Gecko on the stock work with, Javier taking over the feed lot from Pirulo. I felt this would be my time to shine, since being low on manpower would give me a greater opportunity to get more involved with the cattle work and actually be of relevance, as well as also being glued to my radio for important translating duties.

The night before the arrival of the hunters, the *matera* was heaving with excitement. Everyone was double and triple-checking they had everything they needed in their rucksacks: water bottles, rifle stands, binoculars - and the sound of a stag roaring in the distance was enough to give the boys goosebumps. Everyone (well, almost everyone) sat around one table waiting for Ricardo to come and debrief them all on the arrival procedure. While they waited for him, I was sat with everyone, going over a few final pronunciations, words and phrases until, from the darkness, Ruso (who had been one of the chaps thinking he knew enough English to get by and therefore shrugged off most of my lessons) came over in a mad panic, forced aside Chaque who was sat next to me, and took his space. He pulled out a small notebook and a pen and slapped them on the table in front of me.

'Help me, I have no idea what I'm going to say to my hunter!' He opened up his little book and tapped the page anxiously. 'How do you say deer? And what's a boar? Write it down!'

The others were all laughing and I found his hysteria very funny, too. For being so self-confident, Ruso was overcome with nerves and it was totally obvious he regretted his decision to not pay more attention to, or indeed attend, the lessons. As we all teased him about his panic, the look in his eyes was desperate, so I took pity on him and we went through simple vocabulary, where I wrote down translations and phonetics as we went along. He was an intelligent man, so I knew if he just sat down and gave his notes some thought, he would be OK. This half-private tutoring also turned out to be like a test session for the others, who I would get to answer Ruso's questions. In the end I had to stop them though, because he'd get in a fluster worrying about remembering the wrong thing when someone gave an incorrect answer by accident (or sometimes on purpose, just to wind him up even more). More than half an hour later, he had a substantial list. Not a very long one, but enough to keep him going with the basics. 'Don't worry, I'll be on the other end of the radio at all times,' I assured him and all the others.

The following day, the hunters arrived in the early afternoon. While they briefly settled in over at the hotel and left their bags, the gauchos were preparing to meet them and were yet again going through the contents of their rucksacks. There was a lot of anxious waiting as the gauchos were tentative about the introduction to their assigned hunter, with whom they would stick for the duration of their stay. While everyone was packed together at the *matera* whooping with excitement, I looked over across the courtyard to see

Ruso, standing alone away from the hustle and hunched over his little black notebook.

Once introductions had taken place, the boys accompanied the hunters to a 'practice site', where the hunters would try out the rifles to decide which one they liked for the week ahead. As the excited guides headed off, Gillo, Pavo, Gecko, his brother-in-law and I made our way to *La Vigilancia*. This would be my first time up there and my first time meeting the much-spoken of Tío Flaco. We went up by vehicle since a few horses had already been taken up during the week and Tío Flaco had a horse or two to spare as well. We drove for just over half an hour before we pulled up into some cattle pens. Standing inside were some horses and a tall, broad-shouldered man.

Tío Flaco instantly struck me as a very proper gentleman and his appearance was very 'gaucho'. The traditional attire of the gauchos is a shirt worn with a *pañuelo* (neckerchief), *bombachas* (chino-like trousers), a *faja* (like a sort of cummerbund but more of a wrap-around type belt), a *culera* (a leather belt which goes over the *faja*), a knife tucked down the front (or back) of their trousers, *alpargatas* (espadrilles) or long leather riding boots, all naturally finished off with a *boina* on their head. It was nice to see quite a few of the gauchos in traditional dress for work, namely Ruso, Gecko and El Tío. Tío Flaco ticked most of these boxes but lacked the *faja* and *culera*, although he made up for their absence by replacing them with a fat cigar that you would frequently see poking out of the corner of his mouth, sometimes even whilst on a horse.

Being related to Pato meant Tío Flaco also had Indian ancestry. Between the two of them, they let me in on some Indian knowledge - like how to use a snake's shed skin to cure a headache and a few words in one of the many dialects. I absolutely loved this but while Tío Flaco was quite open about his heritage, Pato did not flaunt it quite so much due to the stigma that still exists about the

Native American people and the massacres that occurred during the colonialisation era. They weren't the only Indian descended members on the ranch. Whilst many South Americans don't like the term 'Indian', and it can even be deemed incorrect to call them this, those on the ranch that confided in me their history said that although they may keep quiet about it, they are proud of their ancestry and proud to call themselves Indians. Similar to all colonialisations, the story of Argentina is a sad one and the attacks caused the annihilation of many native people, forcing those who survived to flee deep into Patagonia as their spiritual lands were usurped from them. This is why a large proportion of existing native tribes can be found there and why Juan Alberto Harriet rode all the way to *Chubut* to buy the cattle he would use to start his career. In *La Pampa* and other areas in the country, town names like *General Pico* and *General Acha* exist, which were named after the army generals who 'cleared' areas and had their efforts commended by being rewarded with patches of land.

Tío Flaco introduced himself to me. Although he seemed noticeably shy at the moment of our meeting, he spoke with a deep, confident voice and directed me to the horse I would be borrowing. It was a small, stocky mare. What colour she was I couldn't exactly say, as she was a mix between a bright bay and a strawberry roan, with some overo-type patterning along her right side.

'Paulina is much loved. She's very sweet indeed. Here, let me show you.'
As if I needed to be shown that this little mare - who didn't even run off when I tried to catch her - had an immaculate temperament, Tío Flaco got onto the ground on all fours and crawled under Paulina's tummy whilst she just stood there with my reins crossed over her withers while I stood back and watched.

'I believe you!' I chuckled.

We spent the afternoon moving some cows and separating the bulls. Then we returned back to the *casco* just after sunset. Paulina was a little sweetheart. Despite being short and rotund, her choppy little stride was fast and powerful, helping us keep up with the bulls with no trouble at all. One bull cut away from everyone and ran off. Instantly, Paulina charged after it as far as her little legs could carry us, puffing with every stride, and we managed to turn him around. We kept hot on the bull's heels to make sure he wouldn't try to get away again. Tío Flaco saw us and yelled at us to slow down whilst he laughed, saying we would give the bull a heart attack for making him run so fast in the chase. I hadn't really realised, so I pulled her up and got closer to Tío Flaco, where I apologised to him.

'Don't worry, you're just learning. Besides, I know that because of that little mare's short strides, sometimes you don't realise how fast you're actually going,' he said calmly in his lovely, deep voice.

Returning to the *casco*, I always loved the dusk skies on our way back from *La Vigilancia* as the sun lowered itself behind the mountains at *La Gitana.* Upon arriving at the homestead, the hunting team were sat together chatting about how the afternoon had gone with their hunters and were laughing at how Ruso had been paired with a hunter who was a tall, strong, well-built man. Despite Ruso being built like a beanpole, he was still a fair bit shorter than his hunter. As a result, the jokes were inescapable as the resemblance to Laurel and Hardy was uncanny. Apparently this was a returning group, so some guides were paired with the same hunter from the previous year. Seeing Pato and El Tío's hunters greet them like friends was very pleasing, since they had been paired up together for some years in a row. Despite very limited communication between the American hunters and the Argentinian guides, all the boys said the hunters seemed nice and they were excited for the week ahead.

During the rut, stags roar predominantly when it's dark, so your best shot at finding one is either in the early hours of the morning or at night. You can also apparently get lucky on cooler days, especially when it's overcast. However, I was told this was an unnaturally hot year. By March, although the days are renowned for being warm, it's not unexpected to wake up to a slight frost but that year was totally different. The days were much hotter than they ought to have been and in the morning, one could set off to work in just their t-shirt. This meant the guides had a much harder time seeking out deer, simply because it was still too hot for them to be active. The hunters understood this and were patient – certainly more patient than some of the guides, who desperately wanted to find a beast. Not only to make it worthwhile for their hunters, but also because the boys were competing against one another and every animal spotted, whether shot or not, was a credit to their hunting skills and tracking expertise. The hunting days were long; the party would depart by 5:30am, return for lunch and then set off again at around 4pm until nightfall, or until the hunter had had enough. Every night, the team would always get together with Ricardo as he would debrief them and pass on any messages from the hunters. Also, to avoid accidents, the gauchos would pick a lot to hunt in – one hunter and guide per lot. This was a crucial part of the planning and also demonstrated how well the workers knew the ranch as they would fight over the best lots, knowing where they were most likely to find nice stags and big black-bucks.

Usually, we riders would finish our morning jobs at around 11am whilst the hunters were still out stalking, until they called by radio asking to be picked up once the temperatures rose and the deer went into the shadows to seek shelter from the heat. I remember one morning I went with Pavo and Gecko to collect Enano and his hunter and we would then swing by to collect Pirulo and his hunter along the way. Having collected Enano, as we were on our way to collect Pirulo, one of the tyres on the truck burst and we needed to call for Petaco to come with a spare. Ricardo passed us in his truck and, to keep them from waiting, gave Enano and his hunter a lift back to the hotel, leaving the rest of us to wait for the tyre to arrive. In the meantime, we just had to wait and find a way to entertain ourselves. We had broken down on a path near *Don Juan* on an area of higher ground. I stood in the back of the pickup and looked out across from where the truck sat temporarily dead and could see right the way across to the *pampa* of *Etcheto*. The view was incredible. I happily sat under the sun in the back of the truck until I watched Gecko walk over to one of the trees, pull off a seed pod and start eating it.

'What do they taste like?' I asked curiously.

'They're sweet. Go on, try one,' he replied.

I jumped out of the back of the truck and walked over to the tree closest to me, grabbed a seed pod and bit into it. Immediately, I spat it out. It was really bitter and nasty, leaving a harsh raspiness on my tongue.

'Urgh! That's horrible!' I shouted at Gecko whilst he and Pavo laughed.

Once he finished laughing and doing awful impressions of my disgust, he called me over.

'I'll show you the difference,' and he walked towards the tree he picked the seed pod off.

'This is a carob tree, not a *caldén*. That's why the pods are sweet. What you just picked off was a seed pod from a *caldén*,' he added.

I stood looking at the tree in confusion. It looked exactly like a *caldén*. Even the pods looked identical – they were the same size, colour and even hung from the branches all twisted and folded the same way.

'It looks the same. How do you tell the difference?' I asked, totally puzzled.

'The leaves. If you look closely, the carob trees have longer leaves with a bigger gap between each leaflet, which are thinner than the leaflets on the *caldén*.'

Upon comparing the two, I was able to see the difference. The leaves of these two trees were paripinnate – that is to say, like a mimosa or perhaps vaguely even like a fern. The *caldén* had smaller, fatter leaflets located closer to one another and were more of a dark blue-green colour, whereas the carob had longer, thinner, more needle-like leaflets with a bigger gap in between and a much stronger green colour. Once this had been pointed out to me, the difference was really very clear. And so, with this, I grabbed a carob pod and took a bite out of it. It had a slight pea taste to it, as you'd expect from a legume, but this was accompanied by a sweet fleshy sap, almost like honey in flavour and texture.

'Ooh, that's really nice!' I said to Gecko with a big smile on my face. He smiled back as he watched me nibbling away at the pod, although deep down I know he would much rather have seen me bite into a bitter *caldén* pod again.

Finally, some minutes later, Petaco arrived with the spare wheel. Between the three of them they changed it in no time and Petaco jumped back into his truck to return to the *casco*.

'Fancy a lift?' he asked.

I looked over and saw Gecko and Pavo still faffing with putting the tools away.

'Sure!' So I jumped in and waved goodbye to the other two as I abandoned them.

'Ah, we see how it is!' they joked as I drove away.

The following day Gillo, Pavo and I went back to *La Vigilancia* to help Tío Flaco finish a few bits and pieces with the cattle. Once we finished, we brought all our horses back to the *casco*, so I said goodbye to little sweet Paulina, as she would be staying with Tío Flaco. I hopped onto El Tío's tall, slender dark bay that he had taken up, expecting to do some work up there before the beginning of the shooting. Just as we set off, I heard someone whisper on the radio.

'*Sofi Sofi.*'

It was Petaco. I had been summoned!

'*Si?*' I replied, anxious to know how I could be of service.

'230 yards.'

'What?'

'230 yards,' he whispered again.

'What do you mean 230 yards!'

'What's that in metres?'

'Wow, you could have started with that. Umm...' I sat still on my horse, eyes closed trying to do a rough conversion in my head. 'About 210, I think.'

'*Gracias.*' And with that, the radios fell silent again.

To be honest, it wasn't quite the translation job I had expected. I rather thought my job as a translator was to be between English and Spanish, not between imperial and metric measurements. It was a translation of sorts nonetheless, so I was glad to have been called anyway.

Pavo, for some reason, had taken two horses to *La Vigilancia* so on the way back, he rode one and led the other, which happened to be Moncho, the first horse I rode at *San Eduardo* when I went out for a ride with the family on my first morning. However, not long after setting off, he got bored of leading him so he tied the lead rope around Moncho's neck and let him go. I eyed him up wearily but to my surprise, he just followed along. When we walked, he walked and when we cantered, he cantered. It was amazing how he knew Pavo; Moncho would follow beside or behind him at all times, so much so that at one point, whilst Moncho had fallen back to munch on some carobs, I rode up behind Pavo. After his brief snack he trotted over, pushed me out of the way to get in front of me and resumed his place behind his master. A calm, gentle horse of a sweet nature, very much demonstrating an example of 'horse like rider'. It was funny to see this for a lot of the gauchos: Cabeza being the foreman was, therefore, a man of authority and order, but on the other hand he was also quick to snap if a job was not done correctly. He had horses to reflect all these traits: Patonegro and Mala Cara, who were sturdy and respectable, whereas his *rosillo* would tell off other horses for stepping out of order. Gillo, much like Pavo, had calm, gentle horses and Pato, who had grown up at *San Eduardo*, had older horses that were wise like himself, who really knew the terrain. Then there was Ruso, whose each and every horse was as bonkers as he was to the stage that it made you wonder; was he allocated mental horses to match his character, or did they go mad because of him?

As we made our way back we passed a few trucks full of hunters and guides, who were also heading back for lunch. We cantered along and as the trucks drove by, windows dropped and there was a chorus of '*hola mi amor!*' and '*wey hey!*' along with a quieter, more composed 'hello' and 'good morning' coming from the hunters. What I liked most about the hunting were the stories the guides would come back with after each session. One day, Pirulo said he and his hunter were lying at the top of a sand dune trying to track down a large Royal (a stag with 12 points – 6 points on each antler) they had lost sight of. They spent hours looking, waiting, mocking the same sound the deer make when they roar in order to attract his attention and lure him out, but to no avail. In the end, they called to be picked up and when they stood up to walk to the truck, the deer also stood up and ran off. He had been lying at the base of that very same sand dune behind a small *calden* bush, perfectly camouflaged the entire time.

The best stories, almost unsurprisingly, came from Laurel and Hardy, which included not seeing a perfectly hidden stag until they were almost face-to-face with it. However, the best story came from the same day that everybody got lost; it was a cooler, still, overcast day, so all the hunter and guide groups went into the *monte* lots thick with *caldenes* to boost their chance of finding some stags. The gauchos would always use the sun, shadows and wind to navigate the terrain. This way, they wouldn't get lost. However, when you're in thick vegetation and it's cloudy without even a slight breeze, this is not so easy and they all ended up disorientated. Ruso downright denied he ever got lost though. He told this heroic story of how he spotted an incredible stag so he 'hid' his hunter in the thickets and followed after the beast in order to track it down, before returning for his hunter for him to take the shot. However, the shot was missed and the animal got away.

The evening before the hunters left, we all went to the hotel for a buffet-style stand-up meal, giving everyone the chance to have a chat. Needless to say, I was rather busy that night translating for everyone, but it was lovely and got even better when I was stood in a small group of both gauchos and hunters and Ruso's hunter asked me to translate a story.

'Remember a few days ago, when everybody got lost because it was overcast and the air was still?' he started. This sounded familiar.

'My guide got *very* lost too, so to save me from wandering around meaninglessly, he got me to stay in a spot where he would find me again whilst he went off to try and see where we were. Even though he only walked off a few metres and I had him in sight almost the whole time, I was so worried he wouldn't find me again. I took off my white vest, which I wore under my camo top, and threw it up the tree he had made me sit under, like a sort of flag, so he would be able to locate me again.'

Here I heard the real truth and between breaths of air from laughing, I managed to translate the story to the gauchos, who also could not stop laughing. They couldn't wait to tell Ruso that they knew the *real* story and even when they told him they heard it straight from the hunter's mouth, he down right denied it. Moreover, every time the topic was brought up, to his dismay (much like Javier's 'en-cow-nter' at the lagoon that day in *Etcheto*), this story was never forgotten and Ruso did not ever live it down. He went down as the guide who got so lost that his hunter made a make-do flag in fear that he would be left stranded, alone in the vast *Pampa*, forever. What's worse is that he would get so worked up by people teasing him that it was simply irresistible not to. Then again, he was no martyr as he loved to tease everyone else in return.

PANDEMIC STRIKES

Words cannot express how much I enjoyed my life on the ranch. A few days before the first hunters were set to arrive I had been discussing on calls to home about potentially extending my stay by an extra week or so, simply to enjoy a little bit more time there. What happened next, however, was far more grave and gave me an incredible realisation of how karma (or was it fate, perhaps?) can work upon you. It was March 2020, a date that has become hugely significant and still triggers a strange sensation amongst us all: a bizarre amalgamation of fear, anger, desperation and change.

Just before I left Australia, a coronavirus (which we now know all too well as COVID-19) which had broken out in some market in China, had begun to infect people around the world. Although concerned, at first nobody was really worried about this disease as it started off being described like a fluey-cold, apparently highly susceptible to alcohol and sunlight. Needless to say, Australia was perhaps the worst place to be in when this information was released as the attitude was simply, 'Sun and alcohol? We'll be 'roit mate'. However, in no time at all, as the virus started to spread around the globe and the death counter clocked its first victims, the world began to act. Overnight, borders were closed and all flights were grounded. The American hunters promptly left the hotel to get back to *Buenos Aires* and immediately flew back to the States on the first plane they could catch. I later heard through Miguel that, thankfully, they all arrived well and were in good health. However, the consequences of their stay meant the entire ranch had to be isolated for 2 weeks due to the foreign guests: no one entered or left during this time, including the hotel staff that went just to work in the hospitality for the shooting season.

The government in Argentina acted quickly, closing all the country's borders – both internal and external - in an attempt to isolate the few cases that had appeared. In fact, Argentina's reaction was to act hard and fast and for a long time, it worked. When the entire country shut down the 'stay at home' message was brutal and in *La Pampa*, in order to monitor if people were travelling without permission, truckloads of dirt were tipped onto the roads to ensure every town only had one entrance and exit route, which was permanently guarded by police. They even allocated days on which you could go out to the shops and you had to travel everywhere with your ID. If your ID ended in an even number, you could only shop on a date of an even number and if it ended in an odd number, it was the other way around. If you were caught out, the police would either confiscate your vehicle or fine you what was pretty much a month's salary. Needless to say, these penalties were definitely enough to ensure people did not go about doing anything they weren't allowed to do. In order to travel on and off the ranch so the workers could return home at the weekends, they were given a signed permit by Sergio every week, which gave them special permission as essential workers to go home and then leave the village on Monday morning. Needless to say, this meant the shooting season had been called off. That one group of hunters was the only group that managed to shoot that year. They were lucky though as they got their full week's-worth.

Taking advantage of a translator present at the hunters' send-off, the gauchos and hunters communicated and thanked each other through me. This meant everybody there was able to chat to one another whilst enjoying a nice cold beer and delicious food which, of course, included meat coming straight off the *parrilla*, all juicy and tender. Unfortunately, despite all this food being served it was difficult to eat as I was being pulled from one place to another to translate. I did not mind at all, but it did mean I left the ceremony starving. The boys felt the same, as they felt a shy about going up to eat. Luckily for us, it was Cabeza's birthday that day and to celebrate, a special lamb *asado* had been cooked back at the *matera* for Cabeza and for those who hadn't been a part of the hunting. Being only a few people, there was loads of meat left so after the send-off, all the boys and I went to the *matera* and began tucking in to the meat that was left keeping warm on a tray over the fire. Without any intention of it being so, this turned into a large social gathering and we all pulled up chairs and sat in the shape of a horseshoe facing the fire, all sharing some beers and eating lamb. A few went to bed but the majority stayed: Chaque, Pato, Petaco, Pirulo, Enano, El Tío and two external chaps; one of whom had been there the entire week and the other just a couple of days, and myself, stayed long into the night. We spoke, laughed and the boys all showed one another the gifts left for them by their hunters: boots, binoculars, waterproof camo jackets and trousers and a whole lot more.

The entertainment really kicked off when Petaco, who had rather inconspicuously been avoiding sharing his beers unlike the rest of us, was hit by a tsunami of Dutch courage and began telling us all sorts of silly stories and showing off his finest English. He had the entire room absolutely howling with laughter as he stood before us like some stand-up comedian, beer in hand, and everyone's full attention fixed on him. Just like that, we were telling stories and jokes, laughing away. The hours flew by and when someone eventually checked the time, we decided at 4am that perhaps it was time for us all to go to sleep. Some of the boys had to be at the hotel at 7 o'clock to wave the hunters away but thankfully, being a Saturday, it was going to be a quiet day for the rest of us. For the first time, I decided I would sleep in.

'I doubt I'll be able to function if I meet you all here in 3 hours' time!' I joked to Pirulo and Enano who were standing with me outside as we vacated the *matera*.

'You joke,' added Pirulo, 'but after all these weeks of being up at 6, I can almost guarantee you'll be up at that time. That's what happens when I'm at home and here, I don't even need an alarm clock.'

'Rubbish, I *am* your alarm clock! You're not even up at 6 when you're here, *boludo*!' butted in Enano.

'Trust me, I doubt that,' I replied. 'Without an alarm I could sleep forever.'

I was right. The following morning I woke up at 10:30 and, with a coffee in hand, I went to meet a group of lads who were already skinning and butchering a carcass from the previous day at the butchery. I sat with them, sipping my coffee, when Petaco appeared looking a little worse for wear. He plonked himself down on the ground next to me.

'You all right?' I asked him, to which he replied with a grunt.

The fun and laughter helped numb the uncertainty of the pandemic while it lasted, but it didn't eliminate it completely and the fear would once again hit as soon as any distraction had passed.

These were strange times and the panic, especially back home, was increasing in intensity each day: *When will she get home? How will she get home? Will she get home?*

In the beginning, my mother registered me and my location with the British Embassy in *Buenos Aires*, in the event of repatriation back to England. But, before this was a constant battle with British Airways for flights back. The lack of phone signal and often unreliable internet meant my mother was trying to sort everything out for me, but British Airways refused to speak to her due to 'data protection'. During one of the *siestas*, Petaco drove me to a spot an hour away from the ranch where I could obtain phone signal to call British Airways, seeing as they did not reply to any emails. However, it was no surprise that the phone lines were all clogged up so I didn't manage to get through to anyone despite spending hours trying. I apologised to Petaco profusely for wasting his time, but he reassured me that he was happy doing anything he could to help me. My original flight was booked for the very end of March and with this date fast approaching, the airline still hadn't said whether or not my flight was cancelled. As the days drew nearer, both Britain and Argentina had announced the official lockdown. The panic level elevated yet again and the frequent phone calls between my worried parents and me constantly bounced back and forth between 'No, best you stay there. You could get ill on the flight, bring it home and make us all ill. Plus, the risk of you getting ill there is basically zero' and 'No, you must come home. What if something happens to us and you can't get back? How will we ever know when you can get back otherwise?' In the end, we decided it would be better for me to risk the flight and try to get home. Eventually, my mother had managed to get through to British Airways to get me on their last flight out, a few days earlier than my original flight should have been.

I arrived from work one lunchtime to turn on my computer and found masses of emails and messages from my mother, who was stressed and exasperated but proud of finally achieving a flight back for me after trying so hard. I was relieved too, but also a little sad – relieved that the panic and uncertainty would be over and I would be back home but when I looked at the date in my calendar, I saw I had just over a week left at this amazing place. What could I do? I would just have to make the very most of the little time I had left. After checking through and replying to the emails, I went and sat outside in the sun with my friend Luna, Ruso's puppy that I'd befriended upon looking after her during his weekends in town.

On one weekend, not long after I had arrived, Pirulo, Gillo and I were sat outside Pirulo's room drinking *mate* when I saw a little, black, short-haired Lurcher-type dog under Pavo and Ruso's hut.

'Whose dog is that?' I asked.

'My brother's,' responded Gillo.

I looked at this little dog, who was looking at me with sad eyes as if upset that she couldn't be a part of the fun. Throughout the duration of my travels, the comfort of a canine cuddle from my own dogs was something I really missed, but I was lucky to have had company in the form of a dog on all the farms, including plenty of dogs on the ranch. Her little face got the best of me. Upon asking Gillo to ask his brother whether he'd be happy for me to watch over her at weekends, I got the go-ahead and made myself a loyal companion, to the point where if I left her untied at night, she'd be there waiting for me when I opened my door in the morning, jumping up on me and wagging her tail. She'd then sit under the table on the patio whilst I had my breakfast and would then follow me back to Pavo and Ruso's hut before I went and sat at the *matera*, so I could tie her up to make sure she didn't get up to mischief while I was working.

She also paid more attention to me than to her owner, which although I found this quite funny, it frustrated Ruso greatly.

One night, just as I was about to get into bed I got a message from Ruso asking if I'd seen her, to which I replied saying I hadn't. This, in consequence, meant she had gone missing. He was in a mad panic, searching everywhere around the *casco* for her in the pitch dark. Even with a torch, looking for a small black dog at night was no easy task so I offered my help to him as I looked out of my window and, in the distance, saw his light swaying anxiously in search for her. I quickly pulled on some warm clothes and headed out to help him look for her, hoping that the calls from her owner and her new best friend would entice her back - but nothing. He searched one side of the *casco* whilst I searched the field with the sheep directly behind my room, as I knew she liked being with them for some reason. In some cases, she would even accompany me at the weekends if I went to let them out of their enclosure (they were shut in at night to protect them from pumas). As I called for Luna and shone my torch around the sheep enclosure, the woollies stared at me looking scared and confused, but no little dog was anywhere to be seen. I couldn't even spot any paw prints in the sand. I continued my way around until I met Ruso, who thanked me for my help and told me to get out of the cold, get some rest, and that he would keep searching for his puppy. Even in the dark, I could see the stress and anxiety on his face, worried sick for his girl who was wandering around alone, wherever she was, with the grumpy buffalo and prowling pumas. That was the other thing; usually she would roam around with Tommy, the old ranch dog who was very wise in the ways of the wilderness, but he was tucked away asleep on a pile of wool in the saddlery. Reluctantly, I left the search party and Ruso went off again, continuing his anxious search. God forbid that he ever had to admit it, but he loved that dog. Some nights, from the *matera*, you could see him sat on the steps of his hut with her up on her hind legs propped up against his knee whilst he stroked her.

The next morning I rushed to his hut hoping to find her tied up there, but the lead lay on the ground with no happy little dog at the end of it. I checked the saddlery and looked back around my room just in case, but still no sign of her. Miraculously, two days later on one of his checks, it was Chaque who had found her wandering around looking a little thin, but totally safe and unharmed without a scratch on her. She was in a lot about 15 kilometres away from the *casco*, almost in the direction of *Las Nutrias*. Chaque put her in the back of his truck, brought her back to the *casco* and tied her up at her hut, where her massively overwhelmed owner got to his knees and greeted her the minute he got back from work.

As I sat in the sun stroking Luna's soft, silky tummy, I thought about how I was going to break the news to the gauchos about having to leave early. As I did so, a few minutes later I received an email. My heart pounded when I looked at the subject line: *'Flight Update Notification – Cancelled'*. I called my mother instantly, who had clearly also just received the notification and was yet again in a mad fluster. We shouted – not really at one another, but due to our frustration with the airline since this was the last flight out of the country and they had gone and cancelled it. I distinctly remember my mother's exact sad, exasperated words when I asked her what was going to happen and when I would get back home.

'You are staying, my darling.'

I then remember the strangest feeling fall upon me, which literally reflected the phrase 'my heart sank'. The sadness in her voice, combined with the feeling of *really* not knowing when I would be home, was overwhelming.

'At least we know you are safe there and are loving it. Live your time out there to the maximum.'

She was right. I had no other option. She had done everything she could but the world had been locked down due to this Godforsaken virus and there was nothing we could do to change that. After hanging up, I sat there with Luna for a while longer allowing myself to let the news sink in properly. My poor parents, but at least on the plus side, my mother was right. There I was safe and having the time of my life. I was missing my parents and I so desperately wanted them to visit *San Eduardo* and meet all my new friends, but that was now out of the question and what was even more unsettling was that I had no idea when I would be reunited with them at all.

With both a heavy heart and a spring in my step, I went to the *matera* to meet the gauchos before setting off for work. As soon as I walked in they noticed the strange look on my face - that of strong mixed feelings.

'*Qué pasó?*' they asked. I took a place on the bench and Cabeza passed me a *mate*.

'I'm staying.'

'Until when?'

'Indefinitely,' I replied.

There was a small cheer from everyone at the table and I felt a strong hand on my shoulder from Pato.

'You're teasing! How can you not know?' yelled Cabeza.

I explained how all borders had shut and that my only hope of getting back home had been cancelled. Anyone who walked into the *matera* was greeted with 'Sofi is staying!' I was very humbled by this response. I got on so well with these people, and they with me. It seemed that despite not really knowing what to make of the Spanish-speaking, country-loving foreigner initially, they enjoyed my company. They truly had become my friends, and I had become theirs.

So, there we had it. I was to stay on this ranch for the time being, however long that was going to be. 2 weeks? 1 month? A year? Who knew. All I knew was that I was sure that as time went on, it would get better and better. I had the opportunity to do more, see more, learn more and as we got to know one another better, my ever-growing relationship with the family and the gauchos became more and more special. Despite her stresses, this was at least one good thing that set my mother's mind at ease, since she couldn't have been happier or more reassured knowing I was having an amazing time and how well looked after I was. She was frequently in touch with Miguel, of course, so she was able to communicate her appreciation directly to him. However, since she had no way of getting in touch with the gauchos, she had to thank them through me. Therefore, she recorded an audio message expressing her gratitude to them and she asked that I play this message to them all. So, one day after lunch, once everyone had finished and everything had been tidied up and the floor swept, I gathered all the gauchos in the *matera* and played them the message. They all stood listening in closely, laughing at some parts whilst 'aw'-ing at others and to prove to my mother that I had done as she asked, I sent her a video back of them all whooping, cheering and thanking her for her appreciation.

The family had returned to *Buenos Aires* after the hunters left but being unable to bear the thought of spending the lockdown trapped in their place in the city, they made a break for *San Eduardo* before the lockdown was officially enforced. As for the workers, life went on. With no more shooting to guide, they resumed their normal work. I had really enjoyed the brief hunting season: meeting the hunters, learning about the shooting side of the business and increasing my limited knowledge of butchery to more exotic species. This was one thing I really enjoyed, as I loved the respect these people had for the animals. Firstly, they were a very sure shot. The animal would be taken down painlessly in one go. In other words, it quite literally didn't know what hit it. It was then brought back to the butchery and cold room to be skinned, butchered, and some of it hung. All the meat was eaten by the family and the workers: venison, wild boar, black-buck, buffalo - even the vizcachas. I wouldn't have thought of eating these but they were surprisingly yummy, often eaten in *escabeche* or made into a schnitzel, covered in herby breadcrumbs and then fried. The meat was pale, moist and tender, and it probably won't surprise you when I say it tasted a bit like chicken, just because everything exotic seems to.

The scrappy bits of meat and off-cuts of any animal butchered weren't wasted, as everything was minced up or made into sausages. Before it was, however, I would steal some and take it to feed little Luna and the other dogs. Besides the meat, quite a few skins were kept too: buffalo and cow hides were predominantly kept by the boys and used to make leather goods, given that they made all their own bridles, lassos, *rebenques*, headcollars and other bits like stirrup leathers, reins, or even just keeping the leather to use for repairs on their *recados* or belts. If a lamb was killed, there would be a huge argument over who called dibs on the skin in order to dry it and use it as a saddle cushion. I loved this and I would frequently pop by Pirulo, Enano and Javier's dorm to see what they were up to. Their room faced the round pen at the *casco* and they had a few chairs outside their door where they'd sit, share *mate* and watch the horses as they made their own stuff. I distinctly remember Enano working on a new lasso after his had snapped. The lassos have a specific type of plait that finishes the rope in a squared-off shape, and watching him do it was absolutely incredible. Then there was

Pirulo, who was on his first attempt at making himself a plaited bridle from scratch.

They sharpened their knives on whetstones (the same knives they would carry in their *fajas* all the time) and used these to cut and trim leather straps to fit. It was fascinating to watch them use their big, sharp knives to intricately cut pieces of leather and produce these beautiful goods. Their knives were used for everything: you would castrate bull calves in the morning, cut up leather for your new bridle before lunch and then eat your lunch - usually without a wash in between. This must be one of the reasons they are built like fortresses, because who needs multivitamin supplements when you can keep your immune system ticking over this way? These people were incredibly malleable and nothing much seemed to stop them. Thankfully I didn't witness anyone falling off, but from what I saw on the videos they showed me (often of themselves), they could fall off a horse head-first and they'd get straight up onto their feet, dust off their *boinas* and remount immediately as if nothing happened.

Since there were no hunters, one of the 'jobs' that became available was deer management and culling of bad-blooded and old animals. One evening, Ricardo approached me and asked if I would like to join him on a hunt and naturally, I said yes. Living in rural England I've been involved with pheasant shoots, but I had never been deer stalking before. In the evening, just as the sun started to set, we set off to a place where Ricardo had located a 'chuzo' stag earlier that day. A *chuzo* in Spanish means 'icicle' but in this instance it refers to a stag whose antlers, instead of forming a nice crown at the top, end in a single sharp point. Thus, like an icicle. This is undesirable because in fights, they can easily spear their opponent, either to death or by injuring them badly (usually resulting in a slow death), so for management's sake, these are the first to be culled and since it's a genetic trait, a *chuzo* will always be a *chuzo*.

We parked up and climbed a tall dune. The view was simply gorgeous – the yellow evening sun reflected off the lagoon (the same lagoon that had been used in the filming of the promotional video), which outlined the surrounding Pampas grass at the water's edge with a golden shimmer. The air was warm and the only sounds were those of birds and insects: grasshoppers creaking, the green parrots screeching excitedly and the shrill call of the falcons circling overhead. Even before the lockdown the ranch was never tarnished by the noise of cars and planes. It was only ever blessed with the sounds of wildlife. We perched at the top of the dune for a while, me taking in the beautiful view whilst Ricardo scouted the area with his binoculars before we descended, heading towards a *caldén* thicket, which was where Ricardo presumed the deer would be at this time. He was right. As the sun began setting whilst we approached the thicket, we heard the first roar. As the daylight waned, the deer became more active. Very slowly we crept closer, sometimes crawling on our hands and knees to stay at the same height as the *olivillo*, which was a prickly endeavour. With *rosetas* everywhere, each time you stopped behind a bush you would be pulling these dreadful woody barbs out of your hands and knees. These abominations were so vicious that they would even stick to my leather boots. Ricardo continued to creep closer until eventually, he reached a small *caldén* bush to hide behind. He lifted his hand, signalling me to stop. There was another bush just slightly behind, so I perched there to watch attentively, listen, and pull *rosetas* out of my palms.

We waited a little while. As the temperature began to drop I unrolled my shirt sleeves and it was obvious the cool evening air was having an effect on the stags since the roaring became louder and stronger. Whilst poised I looked around me, observing the *pampa* from this new level. I'd only really seen it all from the elevation of a horse's back, so it was actually rather interesting to experience it from this perspective, down low through the *olivillo* and tussocks of grass; an armadillo's-eye view, if you like. The setting sun was now creating strong golden beams which shot up into the sky, filtering through the clouds whilst the lengthening shadows slowly darkened the ground, giving a whole new character to the place. The auspicious light tinted the pampas grass's creamy-white fluffy seed heads gold, whilst also gently shimmering its way through the silver *olivillo* leaves. With all this and the silhouette of the trees against the golden background, I couldn't resist taking a photo. I prayed that the sound of my camera opening up wouldn't make too much noise and give us away to the deer but luckily, I was safe; the family of wild boar noisily rooting around in the *salitral* right beside us helped maintain our cover.

Bit by bit, it got darker as the sun crouched further behind the undulating sand dunes, and yet the deer still hadn't emerged from the thickets. Ricardo started getting worried since he hadn't brought the night vision sight for the rifle and if it got too dark, he wasn't going to risk the shot. Luckily, moments later, from out of the *caldén* thicket, our target appeared: an old 6-pointed *chuzo* – 3 points on each antler, the ideal cull target due to his age and composition. Despite the warm air, the stag waltzed out of the trees with his head held high. I could see, even from this distance, the puff of condensation with every bellow from this noble animal, which denoted to me just how mighty and powerful these creatures are. He remained a bit too far for a clean shot so Ricardo composed himself a little longer, quietly loading up the rifle. As the creature slowly wandered over, Ricardo propped the rifle onto the tripod, brought it to his shoulder and looked down the sight. I put in my ear bungs and was totally absorbed in the moment. I did not even blink and remained very, very still. Not long after, a massive bang broke the silence and the animal dropped at once. I waited for his signal, just to be sure, before standing up and congratulating him on his clean shot. We collected our things and wandered over to locate the fallen beast, where Ricardo looked him over, admired his shot and then looked up to the dimming sky.

'Thank you Lord for this successful hunt,' he spoke, followed by doing the sign of the Cross.

This was a small gesture which caught me off guard but was very touching. It proved that in order to live on the ranch, hunting was essential in order to eat. You couldn't easily pop into town for food - and certainly not now with the pandemic. So, having all this natural food readily available to us was a real blessing. For Ricardo and the gauchos, hunting wasn't a sport and they valued the life of everything out there. Everything in the ecosystem had its part to play and they all understood and respected that.

We went to the truck, drove it back to where the old stag fell and loaded it into the back. It was incredibly heavy but we managed. As we hoisted the beast up, it was the first time I got properly close to the musky, earthy scent of the deer. It's a unique smell that cannot be mistaken for anything else; once you know it you will never forget it, and once you've smelt it you will be more receptive to it. I certainly was and from that moment, when we rode anywhere where deer had been I'd pick up the scent immediately.

The carcass went straight to the butchery to be processed, some for dinner the following night and the rest frozen for another time. As you can imagine, during the hunting season a lot of animals will go through the butchery, surplus to the hotel and workers. Therefore, the workers could take whatever meat they wanted with them back home for their families, to prevent any from being wasted.

I thanked Ricardo profusely for inviting me along. I thought, as far as things go, it was a lovely first hunting experience... if one can say that, which I now think you can.

THE CEMETERY

As life returned to normal on the ranch with the absence of the hunters, everyone was assigned back to work. We, the riders, were split up into two different groups where the weekend team stuck together, so I went with Gillo, El Tío and Ruso to work in *Etcheto*. After having had a few weeks off, I was reunited with my lovely Patonegro, who was ready for some work again. The four of us had a great time and this was part of what made the whole experience so lovely and why the job was so relaxed and sociable. I loved it as there was always enough time to chat and for me to get to know everyone. We rode and laughed, talking about silly things whilst travelling between jobs. Gillo and El Tío would tell funny stories, which would be interrupted by the crazy Ruso, who would spot a herd of antelope ahead and charge at them, trying to keep up until the animals would prance over the fence and get away. 'Did you get that on camera?' he'd always ask and then get upset when I always said no, because it all happened so fast.

The four of us had many good days out together, including some particularly beautiful days where, from morning to evening, we were met with some lovely landscapes and skies. One evening, we returned to Lot 13 to move the 1,074 cows to another pasture. We started the drive just before sunset and by the end, the sky was bright orange and outlining the clouds with a vibrant scarlet sheen. The cows made their way quietly along the fence line and out of the gate. I stood atop a dune watching this long stretch of cattle walk away and disperse into their new pasture beneath the pink and mauve sky opposite the setting sun. At the end of the day, the horses took their place in the little paddock with the windmill and we were collected just before it got dark.

The following day we worked in Lot 35, which was actually divided into two. One half of it was sown with alfalfa whilst the other half was normal pasture. I hadn't worked in the pasture half and this bit presented itself with all sorts of little majesties. There were patches of a type of grass I hadn't seen elsewhere on the ranch. It wasn't very tall and produced exceptionally fluffy white seed heads, which made it look like you were riding on a cloud. I turned to look behind me and Patonegro and we had left a perfect path. The place was also heaving with black-bucks. It was there where I got closest to these creatures as I came upon a group of them, 3 males and 4 females, drinking from one of the little lagoons. I approached them and got fairly close until they spooked and bolted off, leaping with effortless grace as they went.

This ranch was full of quirks. About half-way through I heard someone on the radio ask where we were.

'We're in The Cemetery,' replied Ruso.

The Cemetery? I could just about guess why it was called this, as the name is fairly self-explanatory, but I wanted to know *exactly* why. What happened there to grant it its name? Was it macabre or peaceful? What exactly was buried there? How old were they? When we regrouped, I asked the boys and they told me that there were a couple of human graves in that field somewhere. I was desperate to see them but I didn't come across them and I wasn't going to waste time looking for them when I had no clue of their whereabouts or what they even looked like.

That weekend, I accompanied Pato in the truck to open some gates in preparation for the week ahead. As we drove, I recognised one of the lots we were passing through. It was The Cemetery. Since he had been at *San Eduardo* for so long, I figured he'd know why it was called that and where the graves were, so I dropped the question. Indeed, he did know where the graves were but unfortunately, he didn't know their history. I asked him if he would mind showing me and with that, he turned off the track and began cutting across the field, through patches of the fluffy white grass. After a lengthy, bumpy drive, we arrived at a small square area that was fenced off with the usual post and wire as everywhere else. Pato slid across a pole which allowed us to get into this fenced bit and there, in amongst the overgrown grass, were three graves, closely dug alongside one another. One of the graves was missing its marker entirely, whilst the other two still had theirs; iron crosses with a small metal plaque engraved with the names and dates of the people resting there. The cross on the right was dug deeper into the ground and its plaque was too badly corroded to make out what was written on it, which was desperately sad. The middle cross, however, stood taller and, for some reason, was better preserved. Between the two of us, we were able to make out some of the inscriptions. The poor man buried there was born on Christmas Day and died on Christmas Day. I cannot remember the exact number of years apart, but it was about 50. 'What tremendously horrid luck to be born and die on Christmas Day,' I thought to myself. I asked Pato if he knew any of the history.

'Unfortunately not,' he replied 'my only guess is that perhaps they are the graves of people who used to belong to *La Gitana* or a *puesto* that used to exist between there and *San Eduardo*, given that where we are now is closer to *La Gitana* than to the *casco*. The only person who would know for sure is Miguel's mother.'

Unfortunately, as a result of the quarantine issued due to the pandemic, Miguel's mother was not able to leave *Buenos Aires*. So, unlike originally anticipated, I never got to meet her, which was a huge shame. I could only imagine the vast quantities of anecdotes and incredible stories she would have to tell about her father. And about this ranch too, as I had been told that *San Eduardo* was her favourite place – and who could blame her?

In the evenings at the weekend, we enjoyed more hunting. On Saturday, Cabeza, Pirulo and I drove around looking for deer until eventually we heard roaring and decided to settle there. There were deer everywhere; tens of does and a few lovely stags, all of which we left alone. Simultaneously, we observed these stunning, noble animals whilst we searched and searched for a *chuzo* until we eventually spotted one. Cabeza, Pirulo and I crept and crept until there was a clear shot with no trees between us and this old *chuzo*. The strong smell of deer lingering around us was phenomenal. I stayed back. With Cabeza to my left, Pirulo crept forward to take the shot. The stag, in a sort of ecstasy as a result of all the does, was oblivious to our presence and was literally walking directly towards us. He must have been a little more than 50 metres away before Cabeza said,

'For goodness' sake boy, are you going to shoot it or catch it by hand?'

With this, Pirulo took aim and pulled the trigger. One shot and it was down.

The following evening I accompanied Cabeza, Pirulo and Ruso to the same lot due to the activity that took place there the previous evening. We perched on a little dune, crowned by *olivillo* and thorny bushes as if it were a nest, and we waited. Cabeza and I stayed out of the way whilst Pirulo was up alongside Ruso, steadying the branch they had found to use as a stand for the rifle. If the day before was active, today was even more so. There were stags roaring everywhere and groups of does left, right and centre. It was incredible. Each time Ruso heard a roar, he'd point the gun in its direction but due to the mass of activity, he was spinning around like a broken compass. Eventually, he sighted a target, stuck to it and, like Pirulo the evening before, took it down with one clean shot. I joked to the chaps that I was obviously a lucky charm for them, given that each time I had been out hunting it was a success, which they scoffed at and denied. Pirulo and I went to fetch the truck and yet another beast was loaded up onto it, for the freezer. Those red stags were huge, heavy animals; males reach around 200kg and out there, they were certainly well-fed and between us, we managed to lift the beast into the back of the truck. Luckily there were four of us that day because it was certainly bigger and heavier than the one Pirulo shot the day before. The simplicity of the 'kill to eat' mentality on the ranch was hugely refreshing. One thing is rearing livestock for meat, but this was something else. There were even times when Juli would say to Ricardo 'there's no meat left in the butchery', so he'd set off and come back with something a few hours later.

This attitude really strengthened the link between me, how nature works and how it is intended. I've always considered myself at one with nature, even more so as a farmer, but my time in Argentina took me on a journey far beyond food production. Working in agriculture, I find the changing attitudes of some people absolutely terrifying. Too many people overlook the true brutality of nature and refuse to accept that all creatures on earth must kill to eat, whether it's killing a plant or another animal. Even worse are those people who won't give up meat but buy it from supermarkets because they somehow believe that no animal suffered to produce the meat on those shelves. The detachment between food and the consumer in this day and age is, I believe, one of the scariest issues modern producers face. More people should experience life as I did out there. A week out there hunting for food; sure, it'll either make or break you, but at least it will cure the extreme ignorance of the modern-day consumer. You'll leave with newfound respect for both the animals and the people who work the land... Or you'll pass out at the first carcass you see and vow never to shed blood again. Either way, at least you'll be exposed to the reality and I do believe that you cannot establish a strong argument for yourself unless you throw yourself into the deep end, whatever your argument may be.

In addition, as I said previously, no part of any animal was wasted. Those chaps were true craftsmen and I often enjoyed just sitting with them and watching how they made all their equipment. They were so friendly and welcoming, too. Some evenings, when it was too cold and dark to sit outdoors, they'd invite me inside to sit with them. If I was really lucky, they'd have some wild boar pancetta to snack on that they made themselves, cured in salt and then rubbed in oil and paprika. It was delicious, so delicious in fact that I took this recipe home with me and bought myself a kilo and a half of wild boar belly and made it myself. It wasn't quite as good, but it was a pretty good first attempt at making it, considering I had never cured meats before. Sometimes, we would sit in the dorm for hours after work right up until dinner, by which point we would be so full of *mate* and pancetta that we wouldn't have an appetite for supper but would sit with the others in the *matera* anyway and join in on the conversation.

We would chat with music on in the background and some of the other chaps would pop their heads in from time to time. I remember one evening when Cabeza and Pavo came in and joined us, followed by Ruso who leant against the door without coming inside.

'Why won't you come in?' I asked him.

'No,' snapped Pirulo instantly. 'He knows he's not allowed in here. He comes in, snoops around, takes off with anything he likes the look of and it takes me ages to get it back, so he's barred from entering.'

This made me laugh. Everyone got on well and nobody *actually* stole from one another. It was just banter to wind one another up. Nobody's rooms were ever locked and, during work, phones were left in the *matera*. I never locked my room and the only person who would go in was Juli, who insisted on going in to clean and make my bed, despite always telling her she didn't have to. I felt bad because she worked non-stop all day, but she was adamant to keep me as comfortable as she could. Nobody touched anything except to be annoying, as I experienced first-hand since they would hide one another's unattended phones just to stir someone up. Usually it was poor Pavo, who would go in to boil some water for his *mate* and return to find his phone had gone missing. This was all very funny, until they decided to do it to me. 'You're just overgrown children,' I would say to them and they would often prove my point by giggling away like some mischievous 8-year-olds. Usually, this was Chaque and Pato.

'*Che, Sofi*,' called Pirulo, who when I looked over was focussing fiercely on the length of leather he was slicing into thin lengths. 'Don't you get bored being on the ranch all the time? Not like you have much choice now because of the virus but even so,' he asked.

'No, not at all. I'm having an amazing time and there's always something for me to do. I love it here,' I replied completely honestly.

'Don't you find it irritating you can't go anywhere?' joined Pavo.

'I guess so. There's a lot to see and now I won't be able to, but I suppose it's an excuse to come back and see everything when I can,' I chuckled.

'Well, you're always welcome to come and stay at my parents' house over a weekend. That way you can at least visit the town I live in. It's not a massive house, but we have a spare room with a bed,' said Pirulo shyly, who still wouldn't lift his eyes off the strap of leather he was working on.

'Likewise *amiga*,' butted in Javier. 'My house is small and humble but whilst you're here, it's your house, too. You're welcome to stay whenever you like. You can hang out with my daughter and spend all day riding if you like as I have lots of horses!'

I felt myself going red and as I looked around at everyone, every one of them was nodding gently and I could see that it meant that they were all offering me the exact same thing. Once again, I was left speechless and astounded by their genuine benevolence, which never asked for anything in return. It made me feel bad. I felt as if I had nothing to give them, which was true because I couldn't even leave the ranch to go into town and buy a crate of beer for everyone to enjoy in the evening, at the very least. They were willing to break the strict lockdown rules and smuggle me into their towns to spend the weekend with them. What's more, they were angry that due to the lockdown, all *jineteadas*, town parties, *asados* and dances had been banned, and that they would not be able to take me to any of these things. They wanted so desperately to show me their culture that they are so proud of, but they couldn't. I was actually really upset about this too, as I would have loved to have gone to all these things but figured, just like I had said to them, that returning once the pandemic was over to do all these things was a firm excuse to go back.

On Monday morning, El Tío, Ruso, Gillo and I resumed our work over in *Etcheto* and were met by Pato by late morning, who had returned from the village. Movement restrictions resulting from the pandemic meant you couldn't travel between provinces, but since most of the chaps lived in *La Pampa*, they were allowed to go home. Those who weren't from *La Pampa*, like Chango and Petaco, had a very long stay at the ranch until the borders finally reopened. As the five of us drove this herd of cattle along, something very strange happened next. I thought long and hard about whether to include this next bit in the book, but in the end, I felt it was important to share this story. This place was incredible but it did not mean anything was exempt from the cruel clasp of nature.

As we rode on, pushing the cattle along, one of the bulls was lingering far behind the rest of the herd, trotting clumsily and anxiously looking around, almost as if disorientated, and it took no expert to see why. His pizzle was huge and inflamed, huge to the point that it was as thick as a log and the animal's clumsy gait came as a result of him trying to avoid stepping on his trailing genitals with each stride. Whether it was an infection or a specific disease I don't know, but it's nothing I've ever seen before or even imagined could be possible. I couldn't help but think of how much pain and discomfort this animal must have been in. The boys knew something had to be done immediately so without a moment's hesitation, El Tío and Ruso galloped off after this bull whilst releasing their lassos from their *recados*. Together they roped the bull as he tried to flee, plucking together all his strength to do so, and with one rope around his neck and the other around his back legs, the two gauchos held the bull firmly whilst Pato jumped off his horse and ran over. He put his arm around the bull's neck as if grasping him in a headlock, reached for his knife with his free arm and, in one quick and confident motion, slit the animal's throat.

Instantly, the creature fell to the ground and in just a few seconds, he was gone. I must admit that it was a horribly tragic event and one of the nastiest things I've had to witness, but the gauchos executed the task swiftly and diligently, ending the animal's long-term suffering – for it had been long-term given the state he was in. Unfortunately, when you're responsible for animals it's not always as simple as feeding them and giving them a jab when they get ill. We only wish it were always that easy. Admittedly, this was a very extreme way of dealing with a situation, but out there, it's just what you have to do. Nature is brutal and out in the wilderness it can claim any victim - be it the wildlife or your livestock. The boys regretfully glanced down at the bull as they remounted and wound up their lassos before promptly continuing on so we wouldn't lose the rest of the herd. As I rode past the carcass, I also imparted a sorrowful glance before averting my gaze and thinking onwards to the task at hand. The poor bull was left where he fell to feed the armadillos and the vultures.

I'm sure many of you will be asking, 'Why didn't we call a vet? Why not herd him alone to the pens and try to treat him there?' Well, the answers are simple: Firstly, the vet lived in *Santa Rosa* which was more than 130km away and the chances of him being available and in a position to dash out to help us were slim. All lives are precious, of course, but on that scale it's a lot of hassle for just one animal and the likelihood of him having recovered from an illness so grave was incredibly slim. Even if the infection was cured and he was saved, he would have been damaged for life and infertile for sure. Secondly, we were a good 5km away from the nearest pens, so herding the bull that distance would have been cruel in itself. Walking all that way with the inflamed, sunburnt infection dragging across the hot sand covered in *rosetas* would have been torture for the poor animal. You may argue that he was already doing that before we found him, but the additional stress incurred by our presence and herding him would have been detrimental. In addition, even well-handled, docile animals can become aggressive when angered, stressed, or pained, so the chances of this feral bull losing his temper and trying to fight back would have been high, which would have achieved nothing more than further damage to himself or to one of us or our horses. It was horrible and I'm not trying to tell you otherwise. However, it's important to understand the reality of the work that goes on in farming, the sorts of scenarios that have to be dealt with and all the outcomes you must weigh up when faced with a dilemma.

Thankfully, the happy, bubbly humour of the gauchos soon blew away the sadness of the event that had just occurred. Onwards and upwards as they say; in the farming world, you can't afford to let things like that set you back, especially when you have thousands of other animals to care for. The shenanigans continued and at one stage, Ruso got off his horse to open a gate and as he let go of his mare, she started wandering off. I went and grabbed her reins and then, to annoy him a bit, I ran off with her as he came to remount. Eventually, of course, I gave her back and I'm not sure how exactly it happened but as he got on, he managed to kick my radio off my belt, sending it to the floor. I was now facing a dilemma. I argued that it was his fault that my radio was now on the floor, so he should be the one to get off and pass it back to me. The trouble is, by doing this he would just pick it up and not give it back. But, if I were to get off

Patonegro to get it, he would do what I did and run off with my horse. I decided the latter would have the better outcome, because whilst he could hold my radio hostage all day, there was no way he would do the same with a horse, so I hopped off (keeping hold of a rein) and collected the radio. Of course, the rein was pulled out of my hand and I saw Patonegro trailing behind Ruso's chestnut mare. Instead of even trying to negotiate to get my horse back, I asked if anyone would give me a lift. Just to continue irritating me, Gillo and El Tío found all sorts of excuses: 'No, my horse is too tired to have another rider' and 'no no, my horse isn't laid back enough to risk carrying you', so in the end it was Pato who helped me up onto Vaso Partido's rump. We rode double-donkey for a while until Ruso, just as I anticipated he would, soon got bored of leading Patonegro and handed him back. Pato directed Vaso Partido up alongside him and I was able to jump across.

Further down the line we met Pavo and Enano and they joined in with us shortly after I remounted Patonegro. It was from this point onward that I was enlightened to a new sport. While waiting to be collected at the end of the day, Pato and his uncle found ways to keep themselves entertained until the truck arrived. These two were often up to mischief, poking at one another to wind each other up. This time, it all started when Pato (with an accuracy that never ceased to amaze me) managed to throw a *roseta* down the very small gap between the back of El Tío's neck and his shirt collar, whilst he was sitting on the ground. When El Tío stood up he felt a strong prick on his back and yelped. Once he had located the bastard thing, he pulled it out, removed it from himself and then angrily walked away to pick up a long prickly plant stem which he intended to use as a whip to get his revenge. However, upon turning his back Pato launched something else at him - and so I was introduced to this new sport: cow pat Frisbee, or 'toss the turd' as I later dubbed it. The disk-like shape of the sundried droppings flew rather well and the battle began when the excrement was flung, once again with impeccable precision, hitting El Tío right between the shoulder blades. With 9,000 cattle on the ranch, there was plenty of ammo to go around and until the truck arrived, these two entertained us all with their dung duel whilst the rest of us sensibly kept a safe distance away, whilst also egging them on... It was too funny not to.

As I would always do, when we got back to the *casco* I popped into my room to leave my gaiters and put my radio on charge, as well as run my face under some nice cold water. It was obviously very refreshing, having been out all morning in the sun, but there was also a strict rule that before meals, you had to be clean. The gauchos were very well mannered and there were a couple of rules that were simply unbreakable. These included not using your phone during meals, putting a hat of any kind on a table (you either hang it on the hooks on the wall or leave it on your head, unless you're eating, in which case you shouldn't wear it) and eating without cleaning yourself up. It was unacceptable to sit at the table with a grubby face full of dirt, especially after working in the pens when you would come back absolutely black from all the dust that had been kicked up. Most people wouldn't want to eat with their face covered in sandpaper-like dust anyway, but regardless: the rules were the rules – and it was incredibly rude to break them.

As I opened the door to my room, I noticed a plastic carrier bag on the spare bed. I looked at it curiously, wondering what on Earth it was and who had left it there, since I certainly hadn't. Juli, who was preparing our lunch in her kitchen, must have heard me open my door and she put an end to my questions.

'*Sofi!*' she called in her high-pitched, squeaky voice.

Considering she was the only one who ever entered my room, I assume she knew what the bag contained and who it was from. I wandered through to see her and before I could even ask, she abandoned her large pot of beef stew on the stove and turned to me.

'You set off too early this morning, before the others arrived from town!' she shouted, almost as if scolding me for doing so. 'Tío Flaco bought you some *alfajores* from town and he wanted to give them to you, but you had already left, so he gave them to me and I put them on the bed,' she said excitedly.

The kindness of these people shone through yet again and once more, I felt myself grinning cheesily. *Alfajores* are a typical South American sweet and they are absolutely heavenly. They are two shortbread-type biscuits, traditionally circular, with condensed milk in the middle, then rolled in ground coconut, which sticks to the condensed milk around the middle. For sure, they are diabetes in a bite, but they are just irresistible. With my stomach rumbling I figured I'd have an appetiser before lunch and went back next door to my room to grab one. When I opened the bag, I found nine *alfajores* of different types. These were bigger than the ones Juli would make and they were coated in chocolate: 3 of milk chocolate, 3 of white and 3 of dark chocolate. I had to thank Tío Flaco immediately over the radio as I had no idea when we were due back at *La Vigilancia*. It could be ages before we returned there and I didn't want him to think of me as a rude person, so although it wasn't the most affectionate way of saying thank you, I picked up my radio and called for him. He answered, 'No problem, I hope you enjoy them,' and with that, I removed the wrapper and took a big bite out of this disk of yumminess – and enjoy it I certainly did.

I went back through to sit with Juli until lunch was ready, where I then carried the large pot of beef stew to the *matera* with Pavo's help.

'Enjoying your *alfajores*?' asked Petaco as we walked in. Knowing he would have heard me talking to Tío Flaco over the radio, I thought nothing of it and replied that yes, I certainly was. He was sat with Chango and Chaque, who like naughty boys started giggling away.

'Why, what have I done now...' I asked with a sigh.

'Tío Flaco switches his radio off during lunchtime,' giggled Chaque, who was going red due to the laughter that was brewing up inside him.

'Meaning?'

'It means – he wasn't the one you spoke to,' blurted Chango, trying to keep a straight face.

'Right... so who did I speak to then?' I replied strongly, trying to sound harsh.

They both said nothing and instead turned to look at Petaco, who stood rigid and unsteady in the corner by the fridge, expecting a telling off from me now that his cover had been blown.

'Oh for God's sake,' I muttered, unable to hide a smile as I shook my head. As I turned away and grabbed a plate, all the laughter Chango and Chaque had been trying to contain escaped violently. Naturally, everyone else who was appearing to get their food was asking what the entire ruckus was about and when they were told, they found it hilarious as well and also began adding to the laughter in the room. For my own good, I really had to learn the voice of the boys over the radio... Never mind, I would just have to thank Tío Flaco later – and hope it really is him who answers my call next time.

The following day we were set to return with the horses back to the *casco* and were greeted with a very beautiful morning. It was cooler that day and all of *Etcheto* was enveloped by a morning dew which lingered in the pockets at the bottom of the sand dunes, creating a very mystic landscape with wisps of silver flooding the base of the ochre undulations in the morning sun, making it look like the dunes were hovering on clouds. It seemed that no matter what season, temperature, or weather, the *pampa* was an ever-changing landscape, proudly presenting you with its gift of unique beauty at every moment.

There was something about this particular landscape that morning that suddenly made me think. As I sat there on top of that beautiful horse, overlooking this mystic masterpiece, I suddenly thought, 'Wow, I am the luckiest person in the world. To go with that, with everything happening right now, all things considered, the lockdown was treating me really well. I couldn't even begin to imagine how awful it must have been for anyone living in apartments and cities; unable to go outside, incarcerated in their homes, trying to escape the potentially fatal grasp of the invisible enemy. On the other hand, there I was – living my best life, wild and free in this immense landscape, riding horses and working cattle with amazing people. The things I enjoy doing the most on this planet.

We were due to do more cattle work in the afternoon, but upon releasing Patonegro back at the *casco*, I saw he was getting very thin, despite his recent holiday. I had noticed this when saddling up, seeing that my girth was almost too long for him, which it never was when I used him the first time. I stood, taking a proper look at him in the round pen alongside the other horses and I knew it was definitely time I should ask for a fresh horse. With his heart of gold, that horse would have kept going for as long as I pushed him but this certainly wasn't fair and besides, I would never do that to an animal.

That afternoon, Pato wasn't going to be joining El Tío, Ruso, Javier and I so he lent me Vaso Partido, the handsome bay who gave me a lift when Ruso rode off with Patonegro. He hadn't worked as many days as the other horses so we would be OK for a few days yet.

The horse was a nightmare. He turned out to be the only horse of the 25 I had ridden during my time at the ranch that I really didn't like. Firstly, he was slow. So, *so* infuriatingly slow. I would fall way behind the others at a walk, so I had to ride him constantly at the annoying half-walk, half-trot gait to keep up. Then, aside from being slow, he was also dreadfully clumsy. Perhaps I had been spoilt by Patonegro and the other horses I had ridden who, even at a gallop, would be watching their feet to avoid stepping into holes. This horse on the other hand would trip over fallen branches in his slow, moping walk. He also had the habit of looking around instead of straight ahead, meaning he would trip up on things even more or walk into trees because he was so absorbed looking at whatever was around him that he would ignore my commands. Thank goodness we only had to walk through a scrubby area and that we weren't actually working in *monte*, else this horse would have probably had us both on the floor.

In fact, this horse was so lazy that he was the only one who didn't increase his speed when chasing a cow. All the other horses would get excited about chasing a cow and you would need to hold them up, but this one? Not at all. There I was, kicking and using the slack of my reins against his shoulder in an attempt to gain more speed, but to no avail. I'm not a fan of whips and the horses were so well mannered that I didn't feel the need to use one. As a result, I never rode with a *rebenque*, even when I was offered one, but I will confess that rejecting one for this horse was a bad mistake.

153

Of course, Sod's law meant that we had a really irritating runaway cow during our drive, a pair of them in fact. Two young heifers absolutely refused to pass through the gate to their new pasture. Cattle are highly intelligent creatures, even the young ones, and they become infuriating when they use this intelligence against you – even more so when they combine it with their speed and bad temper. The red Hereford heifer got fed up of being chased by Ruso, El Tío and me, so she just lay down, knowing there is nothing you can do when they decide to do this. You can push and heave as much as you like but you will never be able to get them onto their feet. You can't lasso them either, because if they're adamant about not getting up, then all you'll do is risk hurting their neck or worse. Therefore, Ruso stayed with her and called for his brother for help for when she would, inevitably, eventually get up. The other heifer, a black Angus, had seized the opportunity of having the attention diverted from her and darted off across the lot. El Tío charged off after her and after being reassured by Ruso that he didn't need my help, I went off to help El Tío.

I couldn't believe this horse. There was a cow getting away at full pelt with a horse hot on its heels right ahead of him. This would be enough to get the adrenaline of any other horse going wild... but not him. He steadily cantered behind, no matter how hard I kicked or growled or shouted. Up ahead, El Tío was fighting with the heifer but she persisted in running away at top speed in the opposite direction. He managed to hold her up against the fence line and when I eventually caught up some moments later, I asked El Tío if he had a plan. It was obvious there was not a lot we would be able to do and he wasn't keen on lassoing the animal in case she hurt herself. Out of frustration, she had also become hot with anger and decided to start charging at the horses, making sure she established a good distance between us and herself. We fought a while longer but with her just getting angrier and angrier, it was to no avail. After asking what to do over the radio, we were told just to leave her.

El Tío and I were now miles away from everyone else and instead of pointlessly waiting for us, they had started making their way back to the *casco*. We walked a bit to allow the horses to catch their breath after the fight with the heifer (certainly El Tío's horse was out of breath from the hard work) and once they were settled, we cantered on to catch up with the others. El Tío effortlessly bounded off ahead and whilst I continued kicking and yelling at Vaso Partido, he still made no effort to speed up. I must confess, I did at this point lose my temper. The sun was setting, we were heading home and even *this* didn't influence the horse's speed. I called to El Tío, who pulled up to a halt whilst he waited for me to catch up and I asked to borrow his *rebenque*. Usually – *usually* – when you hold a whip this is another factor that will influence a horse to go faster, even without needing to use it. I made him know I had it by hanging it down his shoulder and giving it a swing alongside his body – from head to tail - without touching him. We once again kicked on into a canter and he slowly made his way along, just as he had all afternoon and not a tiny bit faster.

Like I said, I'm no fan of whips and hate using them, so I did everything I could to give this horse a chance to make the right decision. I kicked and growled whilst swinging the *rebenque* over my head like a lasso, before waving it next to his head, hoping to get some reaction. But still, he did nothing. He didn't even flinch when the strappy end flew past his eyes. El Tío was laughing away at how frustrated I was getting but by then, I really lost my temper. With one last swing of the *rebenque* over my head to gather momentum, I lowered my arm and clouted the beast on the arse with a mighty 'crack'. From the sound alone it must have stung like a bastard, yet to my ultimate surprise all the horses did was jerk his head a bit...

without travelling forwards any faster. I sat there with my eyes wide and not blinking, my whip arm hanging down in utter disbelief. El Tío of course was giggling away frantically, having watched the show come to such an unspectacular ending. There is no point beating up an animal if a smack like that didn't do anything the first time around. Without looking at him, I handed the *rebenque* back to El Tío and admitted humiliating defeat.

Upon returning to the *casco,* I told Pato that his horse was a nuisance and that I would rather do the drives on foot than use him for the week ahead, which I discovered we would be spending at *La Vigilancia.* When he asked why, I told him it was because his horse was *'más lento que el caballo del malo'* - slower than the bad guy's horse, which is a typical Spanish saying to emphasise someone or something really slow. It comes from films and books, predominantly Western films, where the goodies will always be able to outrun the baddies and get away. No matter how hard they ride, it always seems the villains can never gallop fast enough to catch up with the good guys. Anyway, from bad comes good, and although it was to Pato's dismay that I refused Vaso Partido, this sparked the beginning of a new love affair.

On Friday afternoon, horses for the week ahead were being rounded up from the fields they were set free in and brought into the *potrero* at the *casco*, making it easier to catch them ready for work. On Saturday, we had some light work to do, rounding up some calves to bring to the feedlot and Pirulo, who had been able to bring in his horse separately, had the bridle on Zaino Panzón ready to hop on bareback to bring in the other horses from the *potero.*

'Fancy doing it instead?' he asked.

'Sure!' I said, walking over to him.

He must have asked this in jest because he looked rather surprised when I opened the gate and walked over.

'I was joking, don't worry, you don't have to if you don't want to,' he stuttered.

'Why not? It's OK, I've ridden bareback before, even jumped. Besides, I've ridden this little horse of yours before and I know he's a good boy.'

'OK... If you're sure, if not Ruso or I can-'

'Honestly, it's fine! All I'll need is a leg up, as I haven't quite mastered jumping onto the horses like you do.' I had tried this and been successful at times but my lack of agility was rather embarrassing, so I thought it was better to just ask for help to get on board.

I grabbed the reins and the Zaino's mane and Pirulo gave me a leg up. He then opened the gate for me and I wandered off into the *potrero* to find the horses and, just to make a bit of a point, kicked the little horse into a canter. I found the horses, brought them back and returned to a few awestruck faces. I hopped off and handed an open-mouthed Pirulo the reins of his horse.

'*Sos campera vos!*' he exclaimed, which made me smile. This phrase essentially translates to 'you're a proper country girl,' an expression often used when someone achieves something by doing something crazy or dangerous, such as maintaining your seat when a horse starts bucking or wrestling a calf to the ground.

'Which horse should I use then?' I asked Cabeza.

'Pirulo – can she use your horse Por Si Acaso?'

'Of course, he's that one there,' he pointed out to me.

Por si acaso means 'just in case', so when I heard this I asked them, 'just in case of *what*?'

'No, *loca*! The horse - his name is Por Si Acaso.'

I followed Pirulo towards a tall, slender dark bay. He was a handsome horse and, unlike the others, was easy to catch. You'd always have to work hard to catch your horse, cornering them until they couldn't run off anymore, but this one stood quietly as I approached him and slipped the bridle over his ears. He stood calmly and I took a moment to appreciate this horse's kind expression and velvety nose whilst the others were shouting and whistling, chasing their horses and trying to catch them. My moment with this new horse was interrupted by a shrill '*guarda!*' which forced me to turn around suddenly, just in time to watch the massive brown horse at the end of Pirulo's leadrope towering above us on his hind legs before falling over backwards with a large thud, lifting a large cloud of dust.

'What the hell, that horse is insane!' I yelled somewhat nervously. 'Are you seriously going to use that this afternoon?'

'Yes... He does that but he's OK once you're up and riding,' said Pirulo, surprisingly calmly given that the mad beast was getting up and, immediately upon getting to his feet, forcefully pulled back on the rope again.

Pirulo was jerked forward slightly but with the help of Cabeza, who was right beside him, they both reacted lightning fast and wrapped the end of the rope around one of the legs of the feeder in the middle of the pen, just as the horse threw himself to the ground once more. Being sturdily tied up to the post meant he couldn't rear up so high, so he ended up rolling over instead of throwing himself in full force onto his back once more, before getting up onto his feet once again.

They calmed the horse and untied this large, chunky pale chestnut from the post and led him out of the pen quietly. I followed them out, leaving more than enough of a gap between myself and the psychopathic horse. I walked over to the saddlery where Por Si

Acaso stood calmly and still as I saddled him up, keeping one eye on Pirulo's mad horse just in case he tried any more stunts.

'Just mind when you tack him up and get on; if you girth him up too tightly right from the start he might rear up and throw himself backwards, like this one. Saddle him up loosely, walk him around by hand a bit and then do the girth up tighter when you go to get on,' informed Pirulo.

Rather nervously, having seen Pirulo's horse do exactly that, I took his exact advice on board. I never actually had a single issue with Por Si Acaso, but I would much rather be made aware of any vices like that. Whilst I was walking him around in-hand before getting on, I heard yet another thud from where Pirulo's horse, now for the third time, threw himself over, this time whilst fully saddled up. I cannot for one second deny that I was really quite worried about him getting on to his horse. Pirulo had kindly offered me Por Si Acaso and his other horse, the Zaino Panzón, was being ridden by Sergio, who was making the most of a quiet afternoon in the office by coming riding with us. I felt bad because this meant all his horses were being used, leaving him with that lunatic. 'Perhaps if it weren't for Sergio and me, he wouldn't be using that horse at all,' I thought to myself. So, I asked if there was another horse I could ride so Pirulo could have this one back. He insisted he would be OK and with a brave face, he smacked his *recado* to release the dust from it and collected up his reins tightly so the horse's head was turned to face him. As he put his foot in the stirrup, the horse began to spin with nowhere to go, thanks to the position of his head and nimbly, Pirulo jumped on and immediately kicked the horse on to walk.

Incredibly, after just a lap of the horse lines the chestnut psychopath was totally relaxed and by the time we were all mounted up and were heading off, the horse walked with his head long and low. As we left the *casco* together I was alongside Pirulo and I glanced down at this horse's now very dopey expression.

'What did you say he was called?' I asked him.

Incidentally, this was 'El Jubilado'; the horse whose name means 'The Retiree' because of the way he walked with heavy feet and his head low like an old man.

'I don't believe it! This is not the same horse as back in the pens!' and we both laughed.

Por Si Acaso was wonderful and kind. Much like Patonegro, he was nice to ride, quick and responsive. He was easy to navigate and had a very short, bouncy, but comfortable stride. Unlike Patonegro though, he did have his quirks which I came to learn of on later days (he could get fidgety and toss his head in frustration when being stood around too long and once or twice he reared up when I wouldn't allow him to charge off), but he never did anything really bad and not once did he even come close to bowling over when being saddled up. Amazingly, even El Jubilado behaved himself the whole time and actually looked to be working very nicely on the few occasions I glanced over to make sure he still had a rider on him, which was a huge relief.

Despite having been at the ranch for a month by this point, I was still finding differences between the Spanish spoken in Spain and Argentina. Yes, there were many words that were different or used differently (I had been told off by one of them for using what is a perfectly normal word in Spanish because it means something rude in Argentinian) but it went beyond just words. There were full-on sayings and phrases that had totally different meanings.

The day after taking the calves to the feedlot, we helped Gecko at *Las Nutrias* on a hot Sunday afternoon. It was on our way back that Pirulo, El Tío, Ruso and I were riding together and chatting, as we always would when moving between jobs. I think we started talking a bit about *jineteadas* after yesterday's spectacle with Pirulo's psychotic horse but, as always, the conversation had been twisted around and I became the target of teasing, with Pirulo and Ruso saying I wouldn't be able to sit through a horse bucking. I openly admitted that I probably wouldn't be able to hang on the way they do for the *jineteada,* and certainly not bareback, but most of the time I'm able to sit tight and have had some pretty epic victories in the past. Despite all my years of riding, I can count my number of falls on two hands, if I may so brag. Anyway, they wouldn't believe me and like small children, seeing that it was winding me up meant they pushed even further. Fortunately, without even intending to, I disrupted that conversation by allowing them to feed off my subsequent embarrassment.

All I did was defend myself by making a factual remark.

'I've ridden my whole life in an English saddle, which offers no support unlike your *recados*, and its smooth surface makes it easier to slide off. If I can stay on with an English saddle, I'm sure I can stay on with a *recado.*' Factual and to the point, yes?

No. I then had to ruin everything by adding 'riding in English tack taught me how to ride *'como Dios manda'.*' *'Como Dios manda'* is a very common phrase in Spanish used to describe something properly done, or done to a high standard, and it translates to 'As God demands it'.

Immediately after saying this, the three of them began roaring with laughter so hard that El Tío and Pirulo actually started crying. I thought Pirulo was going to slip off his horse or faint, considering it was a long time before he finally took a breath. Naturally, I questioned their reactions and asked them what was so funny about what I had just said.

'So do you frequently ride naked in your country then?' asked Pirulo, still in tears.

'I beg your pardon!' I yelled.

Turns out, *'como Dios manda'* doesn't mean doing things properly at all. In Argentina, it's used in the sense of 'how God brought us to Earth', so by their understanding I had just told them that by riding in English tack, I knew how to ride with no clothes on. Pirulo never let me forget that one and if he ever needed a laugh, he'd frequently bring this up to amuse himself. I was so embarrassed and was left absolutely gobsmacked. I had no idea what to say and I knew that anything I added at that point would only make it worse.

The laughs didn't stop there that evening, though. As the four of us continued on our way, we eventually entered the lot that surrounds the *casco*, which is where a few different herds of buffalo take residence. Inevitably, we bumped into one of these herds, which had five or six females with calves at foot, all guarded by a seriously huge bull with massive horns. We passed by them calmly at a distance and as we did so, the bull puffed himself up and placed himself between us and his herd, with his massive head high and snorting. Usually, the buffalo didn't take much notice of us on the horses but whether our chatting and laughing had disturbed him, this guy did not look too happy for some reason. In order not to spark a confrontation with him, we paid little notice and passed by, still at a walk, when suddenly for no apparent reason the bull dropped his huge head and began to charge. We all had our eyes on him and the minute we saw him start we all kicked the horses on to gallop away, shouting a series of *'wey heys!'* and whooping as we did so. It was all fun and games and if you can't laugh in the face of danger, then you had no place with the gauchos. However, as usual, it was just my luck that the bull had locked eyes on Por Si Acaso and me, deciding it was us he was going to chase. Maybe the bull saw that I didn't look like I was from around there and thought I'd be an easy target.

'*Guarda Sofi!*' yelled Pirulo through laughter and I kicked on harder, looking over my shoulder at the raging buffalo, who thankfully quite quickly backed down and decided we were all far enough away to no longer be a threat. Once again the three men were howling and wheezing about how, of course, it was me that was chased. 'Your face!' they guffawed and as I looked over at El Tío, his face was totally scarlet and I watched it become darker and darker the more he laughed. He never really teased me, as I think he felt pity for me due to the amount of torture I received from all the others on a daily basis, but clearly this was too much for even him to hide and he genuinely looked like he was on the verge of passing out. I was thick-skinned before I arrived to *San Eduardo*, but without a doubt, my skin grew even thicker there. I obviously hadn't realised that whilst being chased my expression was one of such terror, but upon seeing these three hooligans laughing like mad I couldn't help but laugh as well.

'Look at this *loca*,' cackled Pirulo, 'she almost gets pinned by a buffalo and even still all she can do is laugh.'

On Monday, we had to meet Tío Flaco at *La Vigilancia* at around 8:30am, which meant leaving the *casco* before sunrise. With a long ride ahead of us, Gillo, Ruso, Pato, Pavo, El Tío and I tacked up and set off at quarter to 6 in the morning under the morning stars. Throughout the night, we had been narrowly avoided by a large storm that passed by slowly and from the backs of our horses, we watched the lightning illuminate the dark sky on the horizon. It was a cold morning and the sound of bellowing stags could be heard all around us. For about an hour, we rode in the pitch black before the sun started appearing at just before 7am. We witnessed the sunrise as we crossed the dry salt marsh, where the temperature dropped significantly and we all eagerly awaited the sun to rise fully so it could warm us up. We had all wrapped up rather well that morning, apart from Ruso, whose only extra layer was a thin jumper. He complained bitterly about how freezing he was and that he was going to get ill. Thanks to my odd sense of humour, I then found it quite funny when he showed up at the *matera* the next morning with a filthy cold.

As we rode through an area thick with *caldenes*, we were hit with a strong, foul smell; a smell I recognised as coming from something dead. As the smell got stronger, we eventually came to its source and found a decomposing cow. In England, all fallen stock must be taken away to be disposed of, either by the hunt who will feed the carcass to hounds, or be taken away to be rendered down, since it's totally illegal to leave stock decomposing or bury them. There, on the other hand, living in a natural and wild way also means that you die in a natural and wild way, so if you fail to survive for whatever reason, you become part of the food chain and feed the vultures, maybe the pumas and to my surprise – the armadillos. Before I saw them all scuttling out of the rotting carcass into their little holes surrounding it, I did not realise that, besides eating foliage and grubs, they are also carrion feeders, which is why the gauchos do not eat them. This may be a grim detail, but it just goes to highlight the difference between what farmers in the UK have to do compared to what they dealt with out there. Some may think this is cruel, but it's about as natural as you can get. The conditions in Patagonia are harsh; the summers are scorching hot and the winters are bitterly cold. As a result, if you're not fit and strong, you simply won't survive. Cattle farming in places like this isn't about selective breeding – it's about natural selection and survival of the fittest.

After a long ride, the sun had fully risen into the sky and as we made it to Tío Flaco in perfect time, we got straight to work. By the end of the drive, El Tío and I were holding some cattle in a corner when Pato casually showed up holding this funny-looking fruit. He rode over, stopped next to us and took out his knife, which he used to peel back the dirty skin of this thing, revealing white flesh underneath.

'Here, try it,' he said as he passed over the pale, soft fruit which was a similar size and shape to a sweet potato. 'We call it a *papa del monte* (*monte* potato).'

'Hope you're not trying to poison me,' I said jokingly before taking a bite.

It was soft, juicy and much like a dragon fruit in flavour and texture, only not as sweet and with little crunchy seeds like a kiwi. It was strange and left your mouth with a slight raspiness afterwards, just like when you eat an unripe banana, but all in all, it was rather nice.

'If you know what to look for, you could easily live by foraging what grows here. There's a lot,' Pato said, matter of factually. 'Like this *papa*; it grows underground but once you know what the leaves look like, you'll see them about quite a lot.'

He was right. After Gecko had shown me the difference between the carobs and *caldenes*, I had become quite good at telling the difference and would snap up the sweet carobs to chew on whilst riding through the *monte*. Likewise, one of the thorny bushes which grew all over the place, the *piquillín*, bore little red berries like cranberries, but were more bittersweet and tangy. If you wanted to grab *piquillín* berries, you couldn't just swipe the bush as you rode past like with the carob pods, unless you wanted your hand shredded by the pale, fierce, 2-inch long thorns that surrounded the ruby-red fruits. Therefore, you'd have to stop your horse, carefully take your pick and ensure you had loaded up sufficiently so you could munch as you went along. As told by a hunter himself, Cabeza had introduced him to the *piquillín* berries whilst they were out in the bush hunting, but these didn't make him feel so well. A few other hunters had commented the same, but I loved them and would eat any I could get hold of and they never made me feel unwell at all.

Whilst holding the cattle in the corner, the three of us chatted away happily for quite some time, during which we began to wonder where the others had got to. Of course, nothing could be seen through the thick trees, so El Tío took up the search himself and, like a periscope, stood up in his saddle to see if he could spot anything from up high. I laughed at his technique and admitted I had never actually done that myself. Like any young child learning to ride, I had been made to do 'around the world' to work on balance and I did, at one stage, made two or three attempts at vaulting at one of my riding schools, but I had never stood in my saddle.

'Are you brave enough to try?' he asked with a cheeky grin on his face. 'That horse won't do a thing, you'll be fine.'

'Sure, I'll try - let's just hope he doesn't make the most of stretching his legs in a wide-open space!' I joked.

'No, he's calm. We wouldn't encourage you to do it on a horse which could end up hurting you,' replied Pato. He was right, and I really trusted them so I didn't hesitate to stand myself up. Steadily, whilst keeping hold of the reins (just in case, ironically), I slowly brought my legs up and wobbling slightly, got up onto my feet. Por Si Acaso didn't flinch and stood there quietly without fidgeting.

'See? And she says we're the mad ones,' they laughed, smiling strongly at the huge grin across my face.

There was quite a view from up there. I took my time enjoying it until the fun was cut short and I was brought back into the saddle by the mob of cows, who decided it was boring being stood in a corner and tried to make their getaway, sending us chasing after them. Finally, the others showed up with the rest of the cattle and with all the animals penned up, we travelled straight on to the next lot, leaving the horses at the pens to do the second drive after lunch. I finally got the chance to catch up with Tío Flaco, whom I thanked in person for the bag of *alfajores* he had brought me the week before. With a very shy smile, he nodded his head and as we continued riding along, I could tell by the look on his face that he was glad that I was happy.

I had set a GPS tracker app on my phone from the moment we left the *casco* in order to see how far we had ridden. As soon as I got off Por Si Acaso, I stopped the tracking. The result was 35 kilometres in just under 5 hours, which also included the round-up. With my stomach rumbling away, I was glad we would be leaving the second drive until after lunch.

There were an awful lot of cattle to round up at *La Vigilancia* and it wasn't an easy job over there. Very little of it was open *pampa*, meaning it was mostly thick with thorny vegetation. Not just *caldenes*, but bushes and shrubs like the *piquillín* that came up to knee and elbow heights of the horses. I hated these as I felt bad for the horses, but they seemed to manage to traverse these patches totally unscathed. After dismounting, I would check their legs and bellies for thorns and hardly ever find any. On the rare occasions I did, they were never in deep or causing any bleeding, so that was good. Most of the time, you could push through a few thorny branches but occasionally, you would get into a patch so thick you would have to turn around and find another path around. Working in such bad areas was very difficult, especially when chasing cattle.

The cattle would push through the undergrowth and thorns quite happily and get lost in the bushes if you couldn't keep up by horse. I had lost track of cattle a few times, both at *La Vigilancia* and

167

at *Las Nutrias*, which were much alike in their density of trees. It was very frustrating, particularly if you had been chasing the beast for some time. This work made cross-country riding look easy. In cross-country, you have one jump and your horse would either jump it or refuse it. There, you would be dodging trees, shrubs and bushes – sometimes selecting a different path to your horse and being caught by surprise when they whipped away in another direction. There were even times when, if you had a bush in your way and it wasn't too high, then your horse would decide to jump it (something Patonegro was rather prone to doing). It's safe to say that keeping one eye on a cow, the other on where you're going and trying to navigate the horse down a track not knowing exactly whether they'll take the path or jump the adjacent bush certainly keeps you on your toes.

In one particular, incident I was in hot pursuit of a cow that was determined to get away. I just about kept her within sight as she constantly tried to turn around and head in the opposite direction, or tried to sneak into thickets. She went one way, pushing through the bushes whilst Por Si Acaso and I navigated the best way along tracks that wouldn't lead us into the thickets, which would cause us to lose her for sure. Up ahead, we had a moment of hesitation approaching; we cantered on, weaving our way between branches when suddenly, what appeared to be a nice clear path was blocked by a substantial obstacle: a *piquillín* bush, over a metre tall and about half that in width. As Por Si Acaso caught sight of the bush his ears pricked forward and I felt him start backing off and shortening his stride as he measured up the obstacle. As we drew nearer, I also measured it up and found it to be a nice, inviting box shape - not round and spindly like most were. I glanced over at the cow, who noticing that my horse and I were more fixated on the bush than on her, was beginning to sneak away – almost disappearing from view.

RIDE LIKE A GAUCHO

As the *piquillín* bush got closer, Por Si Acaso lifted his head up higher and higher and with each stride, he became more uncertain and rigid in his back as he doubted what to do next. I once again looked over at the cow and the answer was simple: if we stopped, we were going to lose her. I hadn't jumped for years since my own horse became injured but I remembered the feeling and body language of the horses well, so I knew that and if I softened up and abandoned this horse now, I was going to be projected over his head - something I was not going to allow to happen. Besides, they say that riding a horse is like riding a bike in that you never forget how to do it, so surely the same goes for jumping? At the last minute, I sat up, kicked hard and smacked his shoulder with the slack of my reins and with this encouragement the horse launched himself boldly and effortlessly over the bush, flying with such incredible scope that he cleared it by miles. Upon landing on the other side, he continued after the cow as if nothing had happened. In a moment of overwhelming joy, I let out a cheer and gratefully patted my puffing, wonderful horse on the neck with huge enthusiasm, praising him as we pushed on. I had no doubt Por Si Acaso was more than capable of going over the obstacle and he did so with such ease that it made the hefty bush seem meagre under his shadow as he flew over it. If I *had* had any doubts, I would have pulled him up and accepted defeat, rather than do something that could have resulted in me breaking my neck, or in the best case, being flung head first into the treacherously thorny bush. Looking back, I put Por Si Acaso's hesitation once again down to me and my riding technique. Perhaps he felt unsure whether I would be capable of following him through the jump but with my encouragement, it set his mind at ease and he proceeded to take us over it. I so dearly loved this horse and the trust he had in me and I'm pleased to say all this did not take place in vain. We remained hot on the cow's trail and got her to where she had to be, making it 1 − 0 to Sophia and her newly established eventer.

That coming weekend, livestock trucks were due to arrive to take the weaned calves from *La Vigilancia* over to the feedlot back at the *casco*. It would take far too long to herd them up there with the horses, so although this style of ranching remains very traditional, modern methods are sometimes called for. Pato, Ruso, El Tío and I spent a hot, incredibly dusty afternoon weaning calves from their mothers in the pens and the many cows in the small yard kicked up so much sand and dust that there were times you could not see in front of you. A cow could come running directly at me and I wouldn't know until it had knocked me down to the ground I thought to myself at one point, which was a scary thought. The weaning process was all done on foot, using old feed bags tied to the ends of sticks to direct the cattle. The cattle would enter a small holding area a few at a time, where we would then use the flags to split the calves off into one pen and the cows into another. It was horribly noisy, dusty work and by the end of it we were sweating, exhausted, faces totally black with dirt and my eyes stung as if someone had thrown a fistful of sand into my face, which, in essence, is exactly what had happened.

One of the many surprising things I learnt from there was that the calves were far more dangerous to work with than the cows. Some cows were simply feral and bad-tempered and would charge at you but generally, the majority of them were safe to work with. The calves, however, not being used to humans and therefore much more afraid of the movement and noise when in the cattle pens, would frequently lash out with a hasty kick or charge at you mercilessly. On this occasion, there was one particular bull calf (bigger than the rest) who was incredibly foul-tempered and charged at every single one of us, one at a time. He sent me scampering for safety up the fence not once but *three* times that afternoon but luckily, he didn't manage to harm any of us. Despite ticking nearly every other box, including being chased by deadly snakes, kicked by an ostrich and sharing living space with lethal spiders, I had made it out of Australia without a single confrontation from a cow. In Argentina, however, I was kicked three times by calves (twice in the leg and once in the wrist, which drew blood and left behind a small scar which I bear as a memory of the madness), twice by a horse (once on purpose and once by accident) and I had also been chased by various calves, almost pinned by the angry buffalo and head-butted by an angry cow. Nothing too dangerous luckily, and certainly nothing that caused any major harm. I just look at it as good character building, with plenty of amusing stories to go with it.

Thankfully, loading up the calves onto the truck at the weekend was not such hot, dusty work and nobody was sent up fences to escape angry babies. It all went rather well and, as usual, the job was done with good humour and laughter. Every time an animal got near me I would always hear *'guarda Sofi,* mind they don't hurt you.' It was remarkable how much the gauchos looked out for me, even whilst working - *especially* whilst working. They always seemed to have one eye on what was going on and one eye on me, ensuring I was never in a position of risk. We stood, sweating, catching our breath after sending the last truckload of calves away and I glanced over at El Tío, who, through a dripping brow, was looking at me with a sad, almost remorseful, expression on his face.

'What's the matter?' I asked him, very worried there was something wrong.
He looked deeply into my eyes, holding my gaze for a few moments in silence before saying 'I'll really miss you when you're not here anymore, but I'm really pleased you can stay longer.'
All I could do was smile.

'Seriously. To me, you are a friend, a very good friend.'

These words, accompanied by the sincerity in his eyes and in his tone, left me speechless. I was deeply touched and so lost for words, meaning all I could bring myself to do was put my arm around his shoulders and, with a red face, I continued smiling hugely at him as we walked over to the truck to get back to the *casco*.

THE GAUCHIFICATION

Steadily, I was catching the gaucho fever and becoming 'gaucho-fied'. I was happily sharing *mates* with everyone and, bit by bit, I went acquiring the attire. It all started one day when we were sat at the *matera* and I first paid notice to the *faja* Pirulo wore. It was pretty, with nice colours and patterned with the traditional Argentinian diamond – the same one you find on the Argentinian polo belts that have become so popular in England that they practically make up part of the 'countryside look', even to the extent of appearing on dog leads and collars, or on polo equipment like girths and browbands. I learnt that this pattern was called the *'Guarda Pampa'*, which translates to the 'Protector of the *Pampa'*. There are quite a few debates over where it comes from and what it symbolises but most commonly, it is suggested that the *Guarda Pampa* is an insignia created by the Mapuche Indians over 5,000 years ago, symbolising hierarchy and each colour represented also has its own meaning: black signifies nobility, whilst red portrays blood, representing warriors. The pattern itself is symmetrical, which some suggest it like a mountain's reflection in the lakes of its foothills and the tessellation of the pattern is akin to the mighty Andes mountain range.

'What do you call it?' I asked Pirulo as I pointed at the fabric belt wrapped around his waist.

'This?' he pointed to the cummerbund-type item. 'It's called a *faja* and all the best gauchos wear them,' he replied mockingly.

'I like it. That's on my list as something I'd like to get!'
This, as it turns out, was a very foolish thing to say as a couple of weeks later, Pirulo returned to the ranch after his weekend in town and handed me a beautiful black and white woollen *faja*.

'I got you this, since you said you wanted one,' he said, suddenly becoming shy.

'You didn't need to do that!' I replied, now beginning to blush myself because of the great, big smile on my face.

'Maybe not, but I wanted to,' he replied.
I thanked him profusely, gave him a huge hug, and immediately put it in my room to keep it safe so it wouldn't get dirty. I loved it and so the weekend when we rounded up the calves with Sergio (when I rode Por Si Acaso for the first time) I decided to put it on. I had to wear it for the first time out riding and Pirulo had to be there. But, since he was always at the feedlot, this was tricky. Therefore, I didn't miss the opportunity and when I showed up after lunch at the *matera* holding it in my hand, a large smile crossed his face.

'Would you show me how to put it on?' I asked, to which he gladly said yes.

'The trick is to get it on as tightly as is comfortable, that way it won't undo,' he said as he began wrapping the *faja*.

'As long as you don't put your foot on my ribs and pull back like you do when you girth up the horses, go for it,' I joked.

'How's that?' he asked when he finished. 'It looks great on you!'

'Not bad... perhaps a little on the tight side,' I replied, my voice a few octaves higher. 'I might just loosen it a fraction.'

Everyone at the *matera* laughed and commented on how very *'campera'* I looked and how, slowly, I was being absorbed into wearing the traditional gaucho uniform. With me in my lovely new *faja* and Sergio joining us for the first time, we all posed for a group photo once the calves arrived successfully at the feedlot.

'Hang on,' said Ruso just before the photo was taken. 'She's missing the essential look.' He rode over to me and pulled his *boina* off.

'If you're going to pretend, at least do it properly and get rid of that cowboy hat,' he teased.

And that was the first time I wore a *faja* and a *boina* together on a horse. From here on, if I was asked by someone to look after their knife, I would tuck it down my back in my *faja* and Ruso, getting frustrated at always lending me his best *boina* bought me my own, which was a lovely and unexpected surprise. A group of vets, not Claudio (whom you'll meet properly later), would come to vaccinate the cattle and what I didn't know was that when they had been with us on a previous occasion, they had been asked to bring over a few goods from town, so when they came the next time, they gave everyone what they had ordered and the boys paid them back. I stood back as some of the chaps collected a box of *alpargatas* or a new pair of *bombachas* they had requested, meaning I didn't pay any attention to Ruso buying another *boina* until he walked over and slapped it on my head. In his typical fashion, perhaps also simultaneously feeling a bit shy, he didn't say anything. I felt a little confused, as I wasn't sure if this was borrowed or given but eventually, its ownership was confirmed and like Pirulo, Ruso didn't accept a single *peso* for it, meaning I was the proud owner of yet another lovely gift from another of my new friends; this time, a beautiful beige knitted cotton *boina*.

I even tried a pair of *bombachas* one evening but unfortunately, they didn't quite fit me. Pirulo and Ruso both tried to get me to wear *alpargatas* but they never convinced me to ride in them. This is because I have always been brought up to wear 'sturdy, solid boots when riding'. They tried very hard though and at one stage, almost won - if it hadn't been that the pair I was to be given were just a bit too big, which upset Pirulo greatly.

And that is the story of how I started my 'gauchification'.

The work at *La Vigilancia* continued. After all the calves were taken to the feedlot, the cull cows were sorted, branded and those being taken away to market were retagged. The truckers returned, the cows were loaded up and off they went. El Tío was a passenger in one truck in order to direct them to and from the feedlot and *La Vigilancia*, whilst Ruso and Pato were passengers in the other two trucks. With all the calves loaded and the trucks away, just as Cabeza, Tío Flaco and I were closing up all the gates after them, we received a radio call:

'Cabeza Cabeza. One of the trucks has got stuck.'

We went to where they were and this truck was indeed very stuck, bottomed out in the sand. One of the other truckers decided to try and pull it out, but they just ended up with the wheels buried in the sand as well. So, here we all were, with two of the three trucks fully loaded with cattle stuck in the sand. We were left with little option but to call Petaco to come with the tractor to get them out. We weren't exactly close to the *casco* and given the speed of tractors, it was going to take him about half an hour to reach us.

'*Flaco,*' called Cabeza (even the trucker had a nickname, which simply meant 'thin') 'have you got *mate?*'

'Have I got *mate*! Of course!' He opened the passenger door and brought out his *bombilla,* a bag of *mate* leaves, a kettle, a flask of water and a little gas burner. He placed the burner on the ground, which we formed a circle around, and proceeded to wait for Petaco's arrival by sharing the bitter beverage. The *bombilla* was passed around and the driver paused at my turn.

'Rusa, do you drink *mate?*' he asked me. As I mentioned before, Rusa was a nickname I was given due to the colour of my brown/blonde hair and is what this particular trucker always called me.

'I do, thank you,' and I took the *bombilla.*

'*Sos campera vos*' said the driver, to which my friends agreed. I had met this chap a couple of times, as he was always one of the drivers who came when the trucks were called for. He amused me, and I amused him. Like the gauchos when I first arrived, he wasn't too sure about what to make of a girl working with livestock and was visibly apprehensive when he showed up at *Las Nutrias* the first time I met him. He saw me in the pens but quickly complimented my work after he saw me run away laughing from a calf that had turned and started to chase me, which in turn made everybody laugh. He loved the fact that I laughed at it and as a result he softened to me and told me he admired my character and enthusiasm – a statement I was more than touched by. 'Not often do you see girls out with the cattle,' he said.

We sat together enjoying *mate* whilst Pato, El Tío and I threw bits of *caldén* seed pods at one another for our own entertainment until Petaco arrived with the tractor and managed to pull the trucks out one by one.

RIDE LIKE A GAUCHO

It wasn't the first time they got stuck (one of the drivers managed to get stuck every time they came) and it wasn't the first time we sat together drinking the truckers' *mate.* Similarly, on a separate occasion, we loaded calves for the feedlot, but this time from *Etcheto.* Cabeza, Pavo, Javier, Enano and I were left waiting so Cabeza asked for Flaco's kettle and leaves and we sat there together, sharing *mate* until the trucks returned. Even in the cattle yards when working hard, *mate* would be served to everyone. Over in the feedlot, Pirulo once showed up with some to share between us and during the long days of pregnancy testing, there would always be one person who would take time out to pass the drink around, to the point that the person pushing the cattle into the race on horseback would have the reins in one hand and the *bombilla* in the other.

Like I said initially, if you thought the Brits were bad with their tea, the Argies are far worse with their *mate.* If you haven't guessed it by now, they won't be seen anywhere without it. It's drunk at breakfast, morning break, before lunch, after the siesta and again when they get back from work in the evening. One can drink *mate* alone but if someone sits with you, it's the 'done' thing to offer it to those around you. There is also quite a technique to go with preparing and serving this fine beverage: the leaves are poured into the *bombilla* and cold water is poured in, which is then sucked up through the straw and discarded. This is to settle the dust from the leaves and also hydrate the leaves, which helps stop them from scalding when the hot water is added. Then, the hot water (not boiling – it must only be up to 80°C maximum so as not to burn the leaves) is poured in, where the proper way of doing it is to always pour the water in the same spot, meaning most of the leaves on the surface remain dry. Miguel and his family were avid drinkers of *mate* the 'proper' way, whereas the gauchos just poured in the water, not worrying about always refilling it in the same spot. That was another thing; most of the gauchos would pour a bit of sugar in, whereas the family condemned putting sugar in *mate* and drank it alone.

'Oh dear!' exclaimed Miguel one day. 'You've let the *chicos* lead you into their bad habits,' he tutted. I must admit, although I would drink it either way, I did prefer it with a bit of sugar and I found that if you drank a lot of *mate* without adding sugar it could give you heartburn, which wouldn't happen when sweetened up.

If you were the one with the very important job of 'serving' *mate,* you would be the one to pour the water and pass it around to everyone. One person would drink, hand back the *bombilla* and you would once again pour the water and then either drink it yourself or pass it on to another person. As many people as there were present would drink from the same *bombilla* using the same straw and that was perfectly normal. Thank goodness *La Pampa* didn't record any cases of COVID-19 whilst I was there, else we would have all been dead meat and we often joked about this - just because we could.

Often, during the late morning break or after the siesta, *mate* would be accompanied by *tortas fritas*. I wanted to know how to make these and learnt from Petaco first, then from El Tío, who supervised my second attempt. These tasty treats always went down well and each time any were made, the gauchos would hover like vultures so as soon as a *torta* came out of the fat onto a plate, it would be snatched up immediately and eaten, warm and crispy, with a lovely soft centre. The *tortas* were very basic and simple to make, by combining flour, a bit of salt, water and (El Tío's trick), adding one ladle of fat to the mixture, to create a floury dough that was much denser and drier than you would use to make bread. You then roll out the dough and cut it, usually into circles but squares would do, so that they are about 10cm in diameter. You then stick a knife into the middle of each *torta* (to stop them puffing up too much) and cook them in fat until the outside is golden and the centre is moist and fluffy. The fat would be obtained from beef carcasses, where the hard exterior fat would be rendered down and the liquid kept for frying. The little bits of fat left over, the *chicharrón,* were never wasted and the boys would eat this with bread and it was delicious. Yet again – nose-to-tail eating and not a single bit wasted.

I had a huge interest in learning about the Argentinian and Gaucho cultures, their ways of life, traditional foods and their music and dancing. I tried several times to get the boys to dance for me but sadly, they never did. The most I got to see was one evening when Ruso started dancing to music with his dog. It was very entertaining, but not quite what I was after. I really got into their music as well. The gauchos would sit at the *matera* with music playing from their phones and Cabeza had quite a few good tunes which he'd play whenever we went in his truck. Bit by bit, I started recognising songs, singers and genres and, in the end, the music I most enjoyed was the *cuarteto* - the genre Ruso mentioned was from his region, *Córdoba.* This music is very upbeat and cheerful and most commonly features the keyboard, accordion and drums, with some sort of shaker like maracas, to sustain the beat. It was happy music and the singers Ulises Bueno and Damián Córdoba were two I particularly liked. I remember one day, Pirulo asked me if I had heard the song *'Gabriela'* by Ulises Bueno, to which I responded by saying I hadn't.

'No! That's the song which made him famous! You must listen to it.'

That afternoon, Cabeza drove us to our horses and Pirulo came with us. He found *'Gabriela'* in Cabeza's music and played it for me at full volume and got everyone in the vehicle singing along. As it turns out, I really liked the song and every time I listen to it now, it takes me back to that very moment; music blaring and everyone singing at the top of their lungs. Hopefully someday, I'll be able to sing along with them now that I know the lyrics to it.

One thing I was desperately determined to learn was the card game, *truco*. I pursued this and by watching everyone else play, I was able to pin down my tutor and managed to force him to teach me.

After lunchtime, Chango would seldom go and sleep the *siesta*. Instead, he would sit in the *matera*, lying on the bench watching television. It was here that I made my move. I rummaged for the deck of cards in the cupboard and when I found them, I plonked them on the table next to him and demanded he teach me. He wasn't too hard to encourage, since he immediately sat up and began dealing out the cards. We played slowly and he went on explaining how to kill cards, what to say, and when. This had to be the most random, complicated game I had ever played. The jargon, the hierarchy of cards, the points scoring system – I honestly had no clue. We were there for a while and then my second choice tutor, Pirulo, conveniently showed up.

'No, no, no, Chango, by explaining it like that she'll never get it.'
He pushed Chango aside, collected all the cards and put them in hierarchal order in front of me where the 4's were at the bottom of the deck, the Ace of Swords at the top and the Kings in the middle amongst the other cards.

'That's such a random order,' I said.

'No, it's logical. Take a good look, go over it in your head and we'll have a go at playing.'
I looked for a bit and after concluding it was still a very random hierarchy we started again. Pirulo put down a 3, so I put down a 5.

'No, no, no, threes kill fives. Do you have an ace or 7 of Swords, Ace of Clubs or 7 of Golds?'

'What? No...' I replied hesitantly.

'OK... let's try again.' He re-dealt the cards and slapped down an Ace of Cups.

Okay, I thought to myself. The Ace of Cups wasn't strong enough to kill a 3, so it must be quite a low-ranking card. I looked at my hand and set down a King of 12.

'No! Come on, think of the order of cards! The Kings are in the middle. They only kill one another and the very low cards.'

And so this ridiculous game went on like this. A 2 would kill the King but the King could kill 4's, 5's, 6's and 7's but not the 7 of Swords or Ace of Cups. Cabeza then entered and gave it a go at teaching me too - but to no avail, either. I was honestly beginning to think I was just totally stupid at this game, or that they just made up the rules and card hierarchy so that every card I put down would be beaten. Needless to say though, I endured the pain of learning the game, sat reading the rules online at night (and no, they weren't actually making up the hierarchy it pains me to say) and eventually, I got the hang of it slowly. One against one, anyway.

Once I got the hang of 1v1 I tried playing *truco* in a team. My first match was with Petaco against Ruso and Javier. Need I say that right from the beginning, this was a car crash. Anyway, I pulled through and tried my best but Petaco would use phrases that I had no clue what they meant and when I looked over at him he'd wink or raise his eyebrows. At one point, he even blew me a kiss.

'What on earth are you doing?' I asked him, very confused and a little bit shocked.

'I'm signalling to you!'

'What?'

'You haven't been taught the signals?'

'What signals?'

Yes! Just to make the game even worse, there were signals one would use to communicate with their fellow teammates. A wink with the right eye symbolised a different card to winking with your left eye; smiling from one corner of your mouth was different to smiling from the other corner and it turns out he wasn't blowing me a kiss at all. It turns out that puckering your lips is the code for one particular card... Utter chaos. I tried several times to play this game in a team and my play would either make people laugh or cry, although most of the time it made *me* want to hide and cry. It didn't help that they took the absolute mickey out of everything I did and the one time I tried to use the signals, of course it had to be the card that uses the 'kiss' gesture and when I gave the signal they all started laughing, saying that I was trying to distract them by flirting in order to turn their attention away from my poor game. I endured all their teasing and plenty of telling off but I tried and tried, practiced and practiced and – no. I still had not master team *truco* by the time I left and although my 1v1 game improved a little bit, I didn't master that either.

JINETEANDO

The *jineteada*, as I briefly mentioned earlier, is the gaucho version of a bucking bronco. The horses are often young and have been selected because they are deemed unrideable, just because some horses are. Sadly, there are some that simply never give in to a rider, the same way that there are dogs that you can never get the aggression out of, no matter what or how hard you try. So, instead of euthanising the animal just because it can't be ridden, or wasting food on it by just leaving it to eat and live freely until it dies (horses can live well in to their 30's), the gauchos have found a way for these horses to earn their keep, both on ranches and in society.

The horse is tied blindfolded to a post in the middle of a large fenced grass ring. The rider, usually a male but on the rare occasion a female, then enters in their finest traditional attire with their *rebenque* in hand and gets on. The rider sits ready, leaning back with their feet forward and their arm above their head holding their *rebenque* and the reins in their other hand, wrapped around their wrist tightly so they don't slip during the ride. The countdown begins and once the horse is untied and the blindfold comes off, it throws itself in the air in leaps and bounds in the most spectacular display of aggression and power. Unlike the cowboys, the gauchos don't tie a rope around the horse's balls, so what you see is the show of pure, unenhanced infuriation of an unbroken horse, encouraged only by the cries of the crowd, the enthusiastic voice over the PA system and the rider flopping about on top of it. It also means that all types of horse (stallions, geldings and mares), are used.

Now, there are a couple of rules. Of course, rule number 1 is the entire essence of the show: don't fall off. Number 2 is how long you have to stay on for and for this, there are a number of different categories. Category 1 is a bareback ride (yes, totally bareback) where you have to stay on for 8 seconds. From there, as the categories increase up to 5, there is basically an extra bit of tack involved. Therefore, in Category 5, you ride with a full *recado* and you've got to stay on for 12 seconds. If you meet the time limit, a bell goes off and the mounted escorts will trap your horse between theirs and pull you off. There are also other rules, including that you're not allowed to touch the horse with either of your hands (so you can't hold on when you sense you're falling) and if the horse throws itself to the floor (as they often do, usually by bowling over backwards when rearing) and stays down too long, you're disqualified. All this is taken into account by the judges to determine the winner. 10 points is the highest score and the winning rider might achieve this score by providing an even ride with the biggest display and when the horse rears it should be up high, upright and stretched out with its front legs high in the air.

In the end, whether or not you win, the competitors go home with an amazing rush of adrenaline, a metaphorical badge of bravery, (bruises, aches and pains if you fell off) and one hell of a collection of photos, of which the best one usually becomes a profile picture on Facebook as soon as they get home. The boys gladly showed me collections of their photos and the part that made me laugh was that the printed photos from the event photographer would have the rider's full name along with the horse's name. This was presented in the format of THE RIDER vs THE HORSE, which made it sound like the contestants in a mediaeval jousting battle. To make this even better, some of the horses had the most outrageous and amazing names. Those that stood out were 'The Drug Addict', 'The Clumsy', 'The Black Widow', 'The Rival-less', 'The Cold Droplet' and 'The Get Down From There' - all genuine names I saw printed on competition photographs.

Some of the boys even went as far as showing me videos of themselves competing, which I had a hard time watching. It is an incredible sport and I enjoyed watching it on television. I was gutted that I wasn't able to watch a competition in person because of the pandemic, but it does make you very nervous, or at least me, anyway. The boys would always laugh whenever I watched a horse bowl over and land on top of its rider, which caused me to shiver and clench my eyes shut. Thankfully they always got up though, and sometimes even more incredibly they managed to stay on top of the horse as it stood back up and continued until the bell sounded, which was quite simply remarkable. Not all were always lucky though. Of the gauchos at the ranch, just from *jineteada* accidents, one broke his knee and ligaments in his leg, another broke his clavicle and shoulder and two broke their pelvises when the horse fell over on them backwards.

Most will watch this sport and think it's incredible. Others will think it's totally barbaric and cruel to the horse but whatever you think, you cannot deny the gauchos' skill and horsemanship as they ride a wild bucking horse - sometimes completely bareback - and stay on. Also, for what it's worth, these horses work for a maximum of 14 seconds each month. The rest of the time they live freely in fields, just eating and doing what horses do. All things considered, it's not a massively intense life for them. It is embarrassing though, because watching a *jineteada* really shows up people like me, who think that having jumped a fence bareback makes you the absolute master of horse riding.

'Morning all,' I chanted, as I walked into the *matera* as I did every morning.

I sat myself down on the bench and was passed a *mate.* Shortly after, Pirulo came in and bid everyone good morning. I replied but he totally blanked me and went in to boil some water, which I thought was rather strange considering he was always so sweet and friendly. I thought perhaps he simply didn't hear me, so I didn't make anything of it. When he returned with his *mate* and his water I asked him how he was and he glowered at me.

'Is everything OK?' I asked, hesitantly. I was genuinely getting concerned by his behaviour and even the others were looking at him judgementally.

'Don't you know what day it is?' he replied sharply.

'N-No...' I stuttered. 'Your birthday?' I asked, hoping to lighten the tone.

'Check the date.'

I looked at the screen of my phone. 2nd of April, which meant absolutely nothing to me.

'What's the second of April?'

'Come on, don't pretend you don't know. It's Falkland's Day!' Oh. Now I understood. I must confess that prior to my trip to Argentina, I knew very little about the the Falklands Conflict. All I knew was – *don't mention the Falklands.*

'Seriously? I know nothing about the Falklands! Just because I'm English, you're going to be funny about it? Funnily enough I wasn't even around when the conflict happened and I know nothing about it. Are you really going to be in a mood with me for that?' I fired away at him in an attempt at self-defence, but he didn't answer. Instead, he took a sip of his *mate* and ignored me.

'Oh come on, you've played your prank long enough!' scolded Enano, who was sitting next to him and barged him with his shoulder. 'Poor thing. Don't be mean, she hasn't done anything!'
At which point Pirulo looked up at me across the table with a grin and sparkle in his eye.

'I'm only joking! I only wanted to wind you up and see what you'd do. The look on your face though!' He laughed hysterically and as an apology he poured some water and passed me his *mate*. When he got up to grab something to eat, he came round to me and put a firm hand on my shoulder.

Being British is so annoying. For centuries, all they've done is annoy the rest of the world. In Argentina, I was given stick about the Falklands and I get hassled by my Spanish side of the family because of Gibraltar. It's a no-win situation.

During that week at *La Vigilancia* my beloved Por Si Acaso had worked hard and was beginning to get tired and lose weight. Pato had to switch horses, so he said he would bring me one of his from *Don Juan*.

'Remember, nothing slow please,' I half joked, desperately hoping to avoid riding another Vaso Partido.
He parted from us in the morning in order to take his tired horse to *Don Juan* and then after lunch, he returned to *La Vigilancia* on a fresh, gorgeous dark bay called Gusano. He was leading a horse of slender build, this time a chestnut with a tranquil face.

'You wanted something quirkier, here you have him. This is Alazán Viejo (Old Chestnut), but don't let his age deceive you. He's one of the quickest horses I've got.'

I gratefully accepted him, pleased to be giving my dear Por Si Acaso a well-deserved rest. I saddled him up and just before I was about to jump on, Pato stopped me and knelt down at the horse's front feet holding a block of wood and a small, weird sort of knife – like a combination of a miniature axe and a billhook, and proceeded to trim the Alazán's overgrown hooves. Because of the sandy, stoneless terrain, none of the horses were shod as it simply wasn't necessary. However, this meant that with no regular farrier visits, the gauchos had to look after their own horses' feet, which is a vitally important task when keeping horses, wherever in the world you are. As the saying goes in England, 'no feet, no horse'. If the hoof grows too long, it can crack or split very easily and the horse is also more likely to trip up, which when you're going at speed is extremely dangerous for both the horse and the rider. Positioning the knife near the tip of the hoof, Pato gently knocked the blade with the block of wood and slowly made his way around, trimming and shaping his horse's hooves. By the end, all four feet looked perfect, uniform and smooth. Farriers will spend years practicing their trade before they are qualified to do their work but out there, with just these two basic tools, they did the job themselves. I know they weren't shaping metal or hammering nails into the soles of their horse's feet (which without question takes a lot of skill, training and expertise), but what the gauchos managed with a knife and a block of wood was as good as I have ever seen from a farrier with all the clippers and files one could get. When he finished I ran my hand over the edges and they were totally smooth; no chips, cracks, or fragments. He wasn't the only one though; all the boys did a very neat job whenever they had to do this.

Working that evening and the following day, Pato was right about the energy of this horse. At first, he walked along so calmly that his nose almost rubbed the floor, but the moment we began working it was a different story: he was quick, energetic and responsive. When not galloping around he would fidget, cantering almost on the spot with high knees and an arched neck, occasionally tossing his head in the air and lifting his front legs off the ground - but not so far that I could call it a rear. Being slender, not particularly tall and the only horse with a mane (all the others were hogged), he had the appearance of a quarter horse and our shadow in the setting sunlight with me in my cowboy hat meant we looked very Western indeed. It all started well and I really liked him. After a couple of days' work, we finished at *La Vigilancia* and returned to the *casco* with Gillo, where I led Por Si Acaso and released him at *Don Juan* when we passed by on our way. Pato too was going to swap horses (again) and change Gusano for his turbo charged mare, Come Culo. As we entered the region of *Don Juan*, Pato gave me directions to the old *puesto* where we would pen up the horses whilst he went ahead with Gillo to actually find them in the thick *monte.* I passed through a gate alone and Pato headed off in another direction.

'Oh, you can let him off now,' he mentioned, referring to Por Si Acaso.

I had led him the whole way from *La Vigilancia* but seeing as he would probably continue to follow me anyway (and even if he didn't), we were now in the lot where he would be spending his holidays, so I could slip off his headcollar and let him go. I managed to hold Alazán Viejo still for just long enough to do so and right away, Por Si Acaso had a good shake, followed by a big snort of relief. I thought he was going to embrace his freedom and bolt away but instead, he gave a big sigh and just stood there, next to the Alazán and me. As I walked on, he calmly walked along behind us and did so the entire way, only stopping from time to time to lower his head to grab a mouthful of grass before trotting to catch up. I thought this was so sweet. A couple of times, he walked right next to us, his head by my knee and I would scratch him behind the ears as we continued along.

Having no idea where I was, nor how far away from the old *puesto* I was, I occasionally kicked on and cantered a bit and Por Si Acaso trotted on behind whilst Alazán Viejo powerfully moved forward with his short and choppy strides, neck arched and snorting as he went along. Finally, I saw the windmill poking up above the *caldenes* in the distance and relieved I had gone in the right direction, we wandered over and waited by the pens. Neither Pato nor Gillo were there, so just to make sure I was right in waiting there I called them on the radio, to which Pato replied telling me to stay put. Gillo, on the other hand, gave his usual helpful input by expressing how surprised he was that I hadn't got lost – like he always did.

In no hurry to go anywhere I sat quietly atop my orange horse just watching Por Si Acaso wander about, nibbling here and there and seeing we weren't going anywhere, he eventually dropped to his knees and had a good roll. He grunted and kicked out in bliss before getting up, shaking himself off and disappearing in a cloud of his own dust. When he settled down again, I listened to the breeze through the trees, the creaking of the windmill and the *cotorras* screeching when all of a sudden, I was sure I could hear the sound of a bell, like the ones used on cattle in the Alps. I looked around me, wondering if my ears were playing games and changing the sound of a chain clanging in the gentle breeze, but I could see nothing. Gradually, the sound became louder and with it came the sound of hooves and the voices of people when suddenly, out from the *caldenes* came bounding an old dun mare with a bell around her

191

neck, followed by almost a dozen other horses being chased by the two gauchos. Having obviously been through this many times before, the herd gathered themselves in the pen and stood catching their breaths, looking at us with their ears pricked forwards attentively as if asking, 'who's it going to be this time?' I hopped off my horse to shut the gate behind them and while Gillo stayed mounted waiting for Pato to swap over, Pato began to unsaddle Gusano and showered him off before swapping his tack onto Come Culo. This dark bay, despite being small, had quite an attitude, hence her name. When it came to Pato getting on, Gillo sat watching in anticipation with a cheesy grin. Confused, I enquired about his expression.

'Just wait and see. That mare has attitude. If she's decided she's in a bad mood today, Pato will be on the ground arse first the minute he sits in the saddle,' he giggled.

Of course, I never wanted anyone to get hurt, but I found the concept of this quite funny so I remounted and also watched expectantly. Noticing he had a bit of an audience, Pato began to look shifty as the mare spun around him in tight circles whilst he had his left foot in the stirrup and tried to gain enough momentum to swing up his right leg. With a large hop, he overcame her spinning and got aboard. Once he was seated, the mare ceased her stupidity and walked off anxiously, head up and ears forwards, but with no signs of wanting to launch him like a trebuchet. Gillo and I both let off a small huff of disappointment, whilst Pato on the other hand looked both relieved and smug. His bottom would remain unbruised another day.

The three of us rode home together under the setting sun and I watched the colours of the sky change as the air began to cool. Gillo's little *rosillo* was really quite knackered after the work at *La Vigilancia* so he made it back home slowly whereas Pato, on his fresh, crazy horse, was way ahead. I stayed back to keep Gillo company but the closer we got to the *casco*, the more agitated the

Alazán was getting, trotting sideways and puffing, so I kicked on and kept him at a steady canter, which seemed to help a little. He knew well where he was going and as we approached the *casco*, he began bouncing about, excited to get home. I had caught up with Pato and we laughed as we crossed in front of the *matera*, with my horse springing off his back legs and tossing his head about, clearly eager to be dismounted and have his dinner. I thought nothing of it; in fact, I enjoyed his liveliness. All the other horses I had used had been so kind and placid that being back on something quirky was rather nice. Patonegro, who wouldn't hurt a fly, tolerated me repeatedly launching myself into his ribs almost every time I attempted to jump onto him bareback from the ground (as the gauchos are renowned for). Por Si Acaso, who despite having a reputation for rearing over when saddling him up was as good as gold and never even attempted such a stunt with me. Then, there were plenty of others, like El Tío's chunky grey mare, Amarok, who also didn't have a bad bone in her body. Sometimes, you just need these sorts of easy-going horses, whereas other times it's enjoyable to get on something more spirited.

At the weekend, we had odd bits to do as usual, so I saddled up the Alazán for work. Cabeza, Pirulo, Ruso, Enano and I moved calves from one lot to another in the fields around the *casco* and this horse was on absolute fire. There, unlike at *La Vigilancia,* the more open terrain really gave me a taste of the old chap's abilities. Rounding up calves was always fun, as it was often absolute chaos trying to round up hundreds of young cattle who hadn't yet had much experience of being pushed about by horses. Throughout, the old chestnut was excelling in the task. Clearly young at heart, he was crossing the sandy landscape nimbly and with remarkable speed. I got a call on the radio from Pirulo, who said that he was gathering a group of calves that had split up into two and that he would carry on forward. He continued by asking if I could gather up those that had split off and turned around, to which of course I said yes, as I wasn't herding any calves at that point. In a second this horse spun around and was off behind these calves, who were fanning out more and more the farther they ran. Effortlessly, this old horse galloped a circle around them, grouped them together once more and got them back in the right direction where they saw the other calves in the distance and thankfully, decided to follow them.

It wasn't until returning to the *casco* that Alazán Viejo started playing up, but worse this time. As if he knew his work was done when the gate was shut behind the calves, he was ready to go home. Whenever we faced the direction of the *casco*, he immediately arched his head over, chin almost touching his chest, and would start cantering sideways. There's not a lot you can do when a horse starts doing that and if I tried to straighten him up, he just started rearing. I tried and tried just to relax and not to make a fuss, but making a fuss was all he wanted to do. Ruso and Cabeza approached and just told me to get off the horse to avoid being hurt. I, being a bit stubborn but also having learnt to 'never give in to a horse misbehaving, else they'll know how to manipulate you', at first refused to do such a thing but after he planted himself and reared up vertically when I asked him to move on, I decided it wasn't worth the risk and took their advice. Ruso got onto the Alazán and I got onto his horse, which wasn't much saner than the Alazán, but at least it didn't rear up.

The following week with Pato, Pavo, Gillo, Ruso and Enano, I was prepared to saddle the Alazán again, as there were no other horses available for me to use at the time. I was given a few looks and some stern words from Cabeza and Ruso, who had been there at the first performance, but I insisted it would be OK.

It was not okay. In fact, from the moment I got on, this animal misbehaved and was completely intolerable. From the other day when he was responsive, quick and amazing up until the point we had to return home, this time he decided to be a nuisance from the moment I got in the saddle. He wouldn't go where I asked him to without a fight. When we cantered or galloped it took me ages to bring him back and between rare moments of brief calm and good behaviour, he would suddenly throw another tizzy and start rearing up and bunny hopping again. To be honest, he was being so resistant to my commands that I wasn't really able to do the job. All I could do was hold on tight with as little fuss as possible in the hope that he would get bored of being an idiot. One moment I was walking behind five calves who were calmly and steadily walking along the fence line when all of a sudden, he decided he wanted to fulfil his life's dream of becoming a Pogo stick. This scared the calves and sent them all running in different directions. When I tried to gather them up, this horse wasn't having that either and in the end, I needed to radio and ask for help in getting the calves back together and moving along.

It was utter chaos. Despite my patience disappearing rapidly, I sat through it and just got on with it, always expecting another episode each time he settled down. Whilst we moved between lots, Pavo very sweetly rode alongside me a lot of the time, especially through the horse's moments of madness. During one of these moments, the horse's frantic jolting unclipped my radio from my belt but luckily I noticed before too long and turned around to retrace my steps. Enano came with me, occasionally speaking into his radio to see if we could locate mine on the ground. As I had expected, we found my radio in the sand at the point where my horse *had* had one of his ultimate stupid moments. I got off, grabbed the radio and upon getting back on, he threw his head up and hit me in the face, something he had managed to do three times that day, and I was beginning to get sick of it. From the ground he still had that sweet, gentle-natured, old horse expression on his face, but you couldn't let that deceive you. His calm appearance from the ground was clearly just an act.

Anyway, the last straw was when Enano and I managed to find Gillo and Ruso, who had lassoed a getaway calf and were leading it to where the rest of its herd were. Ruso led the calf and Enano helped push it on when it reluctantly fought against the rope, leaving Gillo and me riding side by side when my horse had yet another moment. This time, he went all in, abruptly planting his feet into the ground and rearing up completely vertically. He even went as far as stretching out his front legs to give me that extra 'jolt'. Now he was really was asking for a fight. If I *didn't* fight him, he would bronc and toss his head until he got a response and if I did fight him, he would rear even more aggressively.

'Sofi, get off that horse. He's going to throw himself backwards the way he's behaving,' said Gillo seriously, sounding genuinely concerned.

Having ridden for as long as I can remember, a misbehaving horse doesn't scare me at all. In fact, this was the only place where they actively didn't let me ride anything that wasn't well-mannered because I always seemed to draw the short straw during riding lessons. If anyone was to have a misbehaving horse, it would always be me. Alternatively, if someone's horse misbehaved badly enough to throw off their rider or scare them enough to make them want to get off, I would, for some reason be the one asked to get on it and reprimand it in order for it not to pick up bad habits and learn to get away with bad behaviour. However, I'll admit that the thought of one rearing up and falling back onto me does scare me. Needless to say, I'd had enough and didn't need any encouragement. I kicked my feet out of the stirrups and quickly got off before he managed to squeeze in one more stunt. Admitting defeat, I passed my reins over his head and walked him over, Gillo still by my side, to the others just up ahead.

'What did I tell you about that bloody horse?' remarked Ruso as he saw me approach on foot. To be fair, he was right.

We did a horse swap for the ride back home and exactly like that day in *Etcheto*, right at the beginning, we all swapped horses apart from Enano, who once again managed to stay aboard his own mount. Gillo got onto the Alazán Viejo and gave Renuncia to Ruso and I got onto Amarok, whom Ruso had been riding. I can't deny I felt much better getting onto that little fat mare for a quiet ride home. I felt sad about the Alazán because I really did like him to work with – but only when he behaved, that was. What was even more frustrating was that both times I had to get off of him, he behaved impeccably with both Ruso and Gillo and although he was fine the first time with me, perhaps a bit like Renuncia, he was sensitive to picking up my different riding style and didn't like it. To be fair, after over 20 years of being ridden almost exclusively by Pato, I couldn't blame him for this. Like a lot of old men, he was obviously too old and set in his ways to change at this late stage.

We were (peacefully) making our way back to the *casco* under the evening sun when suddenly Enano's horse decided Amarok was too close, despite being a decent distance away and riding alongside him. So, in order to re-establish his comfort zone, he kicked out at her but missed and got me in the ankle instead. As a result, he got strongly reprimanded by Enano. Luckily, it wasn't a proper kick but more of a 'you're too close' one, so it wasn't strong and meaningful but it's safe to say, it wasn't my day. I had lost my radio, had yet another go at a *jineteada* thanks to the old nag and to top it all off, had been kicked by a horse whilst being on a horse... That takes skill. Stating the obvious here, after that day, I gladly let the old dobbin go and never saw him again. I was invited that evening by Chaque to go and shoot vizcachas but I said to him that, given how my day had gone, I wouldn't be surprised if I ended up shooting myself by accident, so I sat that one out. Instead, we got back and decided to have a game of football, which seemed less likely to cause any major damage.

Whenever I played football with the boys, it was always the highlight of the evening. Old buckets would be laid out as goalposts and the boys would split themselves evenly into teams. I could never resist joining in. Although I'm not a fanatic about watching it on television or anything, I do like to have a kick around, so I started on the field but simply couldn't keep up. The gauchos charged around with incredible speed and they were incredibly nimble players. Therefore, on the rare occasions that I would possess the ball, they would take it off me without so much as brushing my foot, whereas with one another they were far more boisterous as you could perhaps expect. Eventually I got tired of running up and down without touching the ball so I swapped places with Enano, who was in goal. He was a great player and was wasted being stood there, whereas I was just taking up space. I also figured that being in goal could give my team a better chance at winning because the chaps were very conscious of not hurling the ball hard at me, which I found quite sweet. At one stage, Cabeza (who was on the opposite team to me) was in possession of the ball and rapidly advancing towards the goal. Fully absorbed in the heat of the game, he hadn't noticed that Enano and I had swapped position so with an almighty kick, he launched the ball at the goal with all his might – which I stopped with my thigh. My goodness it stung, even through my jeans, and when he realised what he had done his jaw dropped.

'No no no! You can't play like that against the poor girl!' I heard from various players, both on my team and from the opposition.

'I'm so sorry! I didn't realise you were in goal!' grovelled poor Cabeza, who ran over apologising and gave me a massive hug. 'Did I hurt you? Please forgive me!'

I could feel the heat from where the ball had hit me. There I was with a stinging leg, having declined to go shooting with Chaque in order to play football because I thought it'd be less perilous.

I promised Cabeza it was fine and that I had encountered far worse playing football in primary school where at that age, boys only cared about showing off and didn't understand chivalry. This meant we girls would be scattered left, right and centre as they ploughed down the middle of the pitch dribbling the ball. I reassured Cabeza over and over again that I was fine and that there was nothing to forgive. With that, the game resumed. There were a few more attempts at scoring but nothing with any great force and in the end, my team were victorious. I guess the pain was worth it and my plan worked. In fact, it worked so well that for all the games we played whilst I was there – which must have been three or four, I always went in goal and my team always won. Not suggesting any link, of course...

As with everything there, we always had a good laugh playing football, to the stage that by the end of each game my stomach would hurt more than my legs. El Tío was always fun to watch as he'd whiz around everywhere like Sonic the Hedgehog, which always made me laugh due to his quick, tiny little strides. At the same time, however, he would practically dance with the ball with exceptional footwork. Chango, who was perhaps the biggest football fan of them all, was utterly incapable of playing without giggling hysterically with amusement throughout the entire match.

Then there was Petaco, who would somehow always get squashed or knocked over. However, the best matches to watch were those which we played after Chango had watered down the yard. To prevent dust from being kicked up around the buildings, there were sprinklers to wet down the sand but if Chango over-watered the ground or watered too late into the evening, the area would turn into a sort of ice rink.

      The boys would be running around with the ball and slipping all over the place, one moment heroically running towards the goal and the next, they'd be looking up at everyone else from the comfort of the ground with a mouth full of wet sand, having obviously lost the ball in the process, along with most of their dignity. This was even funnier was when one would slip mid-tackle, taking the other one down with him, thus causing a great big pile up, which would make even more players fall to the ground in a heap when they couldn't stop in time and crashed. There were a lot of cats and kittens all over the place who loved eating the scraps in the food waste bin, which would frequently be knocked over by the ball, sending the terrified cats fleeing like in the cartoons, with their fur standing on end in terror. As for my goalkeeping, the gauchos decided to get craftier in order to avoid hitting me; they'd try to narrowly skim the ball past the goal posts or kick it over my head. Some had these measurements nailed down, whereas others did not. Ruso once got so over-excited trying to send the ball over my head that he managed to kick it onto the roof of the *matera*. It was Gecko, who would always come over with his family to eat *asado* with us on Thursdays for dinner, who got up from the bench on the porch and climbed onto the roof, sending the ball back down so we could resume play. Being busy with the match we carried on, leaving Gecko stranded on the roof for a bit until he made his own way back down. He never played with us and usually, if a ball went out of bounds he wouldn't have anything to do with it, but clearly the fun of climbing onto the roof was irresistible. These games went on for however long it was possible to see. We would start in the evening once we had all got back at sunset and would keep going until the spotlight on the barn was no longer strong enough for us to see properly. Sometimes, this would mean one of the lads receiving the ball to their face, which would be the defining instance for calling it a day.

We played football at any time, usually spontaneously. All one would have to do is ask someone else if they fancied a kick around and this would attract the others into playing. During the heat of the day we never played properly, but there were a few days when before or after lunch, Pato, El Tío and I would have a quick kick around just for fun. There was a time when Pato and El Tío were passing the ball to each other under the porch of the *matera,* which wasn't the best idea but they did so anyway. With each kick, the ball gained force.

'Mind the light, Tio...' warned Pato.
Of course, with the next kick, El Tío managed to send the ball straight into the lightbulb, causing it to smash into smithereens.

Anyway, usually we played football on a Thursday evening, as Thursdays were *asado* night, since people would go home on Friday. Juli would prepare a salad for us and the gauchos would cook the meat on the great big *parrilla* they had in the *matera.* The meat was always delicious and they did absolutely nothing special to it in terms of its cooking. They salted the meat well, put it on the *parrilla* and just left it there to cook over hot embers until medium-rare and juicy. Even offal and odd cuts of meat would be plonked onto the grill without special preparation. I tried all sorts of strange, bizarre bits of meat I never would have thought of eating, such as *chinchulín* (beef small intestine which was crispy, greasy and when grilled, would twist into a corkscrew) and cow udder (a very strange, bland thing to eat with a rubbery texture). The udder was OK, but the *chinchulín* I liked.

A lot of cooks will take ages cutting out the valves and vascular system of a heart but on the ranch they just cut it open, put it over the fire and it would be tasty, tender, juicy and delicious. Usually, the *asados* were always beef (or buffalo, during the hunting season), but on special occasions we would have wild boar or lamb from *San Eduardo's* own small flock of sheep. Regardless of the meat, whatever came off the *parrilla* was all delicious. Lamb would only be prepared for special events, such as for Miguel's oldest son, Rodrigo's, birthday. He had come up from *Buenos Aires* with his parents and brother to spend the lockdown at the ranch and his birthday coincided with his stay. Juli prepared a selection of fantastic salads whilst the gauchos cooked the lamb in the *matera*. The family went to eat at the *matera*, which I thought was wonderful. The family didn't segregate the workers from themselves and I thought it was lovely how they would occasionally include the gauchos in things they planned. After the meal, a few games of *truco* couldn't be resisted. I wasn't brave enough to participate the first time but on another special occasion that we had lamb, (*La Pampa* had been in drought for a while and it had rained, so Miguel called for a celebration), I braved it and had a go. Despite the assistance from my teammate Petaco and over-the-shoulder help from Chango, Chaque and Pato, I didn't perform very well. In the end I gave up my place to someone else, who went on to destroy poor Miguel and Rodrigo.

This large community provided a great work/life balance, without even having to leave the ranch. As a result, despite living a secluded life on the ranch, it didn't mean it was a lonely life. I personally know of far lonelier farmers back in England who, despite living on a small farm just outside a village, suffer from a bizarre form of loneliness as a result of their own reclusiveness. Many are overcome by this mindset that they cannot leave the farm because there is always something to do, or the feeling that there is no one worth seeing. This danger in the countryside means more and more farmers are being misled by their own beliefs, adopting this distanced behaviour to the detriment of their own mental health – or worse, the taking of their own lives. With a vast number of farmers working alone every day, they don't even have the company of employees or co-workers to submit to and it's heart breaking. Evenings and down-time at *San Eduardo* were there to be enjoyed, and work was also sociable and cheerful. Then, when the workers went home at the weekend, they made the most of seeing people and having fun, where often they would gather at a *jineteada* before going on to have an *asado* with a large group of friends and family.

LAGOONS & LEISURE TIME

Whilst I went along with a couple of the others on an impromptu trip to *Las Nutrias* to help Gecko with some stock, the rest of the riders went to *Etcheto*. After lunch that Friday, I sat with Cabeza on the porch of the *matera* when he ordered that the group of riders who had been in *Etcheto* bring their horses back before they left for the weekend. This entailed being driven out to the *potrero*, saddling up, moving the cows in the handling pens out to their pasture and then bringing the horses all the way back to the *casco*, which, as you have probably realised by now, was not just a short ride away (it took 25 minutes by car, so around an hour and a bit by horse). What's worse – this meant working through their all-important *siesta*, which was not an order greatly received by any of the boys.

'I'll bring a horse back, I don't sleep the *siesta* anyway and I'm always happy to ride,' I offered enthusiastically.

'No, you don't always have to be on a horse,' replied Cabeza sharply. He teased me quite often but that time, I struggled to make out if he was joking or genuinely saying I couldn't ride.

He called over Pato, who he got to gather everyone up and get the truck. Ruso was one of the people with a horse in *Etcheto*, but he'd managed a timely escape and had already got to his hut and fallen asleep, so I offered to bring his horse for him.

'No,' replied Cabeza sternly again. 'What you can do is go with them and bring the truck back.'

I'll be honest, I wasn't very happy. In fact, I was rather angry. Why wouldn't he just let me ride? I was jumping at the chance, so why was he getting funny about it? Everyone was moaning and groaning about missing their *siesta* and having to work during the hottest part of the day but then I, a willing volunteer, was being told I couldn't ride. With Pato at the wheel, I sat in the passenger seat with El Tío, Gillo and Pavo in the back and Pato laughed at me along the way.

'I've never seen anyone look so angry,' he giggled. 'You have a face like thunder and when Cabeza said no to you, your face went totally red. All you're missing is having some steam come out of your ears,' to which he laughed even more.

We arrived at the *potrero* and everyone hopped out, apart from Pato.

'Well?' I grumbled, 'aren't you going to get out so I can take the truck back?' to which he started laughing again.

'With the mood you're in? Don't be ridiculous. I'll drive back, you go and saddle up,' he smiled.

My expression must have changed quite dramatically because he really set off laughing at this point and did so even more when I thanked him profusely.

'Will Cabeza not have something to say? Will you get into trouble?'

'Don't worry about me. I can handle Cabeza.'

I looked around and spotted Ruso's mare, which I assumed I would be riding as he was the one missing. I walked over to the *caldén* where everyone's tack was hanging from the branches and I began sifting through bridles, searching for Ruso's, before being halted by El Tío, who said he would prefer me to ride a different horse.

'Take Minifalda,' he suggested.

'But that's Pavo's mare' I replied. Pavo was with us, so I didn't want to nick his horse.

'No, that's okay,' replied Pavo, 'you take her and I'll ride the other one, but I want my tack so you use Ruso's'.

Fair enough, so I caught Minifalda who was a pretty, very squarely built dark bay mare – the very typical 'one leg in each corner' sort of horse - sturdy and robust.

'Watch out when you do up her girth because sometimes she kicks,' I heard Pavo say.

With all of us ready, we released the cows from the pens and took them over to their new pasture. Minifalda was cool, but riding in Ruso's *recado* felt really odd. It was more comfortable than mine, *seriously* more comfortable as it was layered with a thick new sheepskin and it was well-moulded, meaning I was held securely in place and felt as if I was sitting in an armchair. It's no wonder how easy it is for the gauchos to sit through a horse bolting and bucking! However, the elaborate wooden stirrups, being made for *alpargatas*, were too small for my boots. Therefore, I was riding with my feet out and the sponges of the *recado* were so thick that I was swaying from side to side as I rode along, which felt as if my saddle was going to slip. Concerned about this, I leant down and checked the girth, which was a fraction too loose, so once the cows had passed through the gate, I dismounted and began sorting myself out.

I undid the whole *recado* and corrected it all whilst Minifalda stood there calmly, until it came to doing the girth. Pavo approached and literally as he said *'Guarda Sofi*, remember I told you she kicks,' she lifted her left hind leg and kicked forwards, catching my shin with the tip of her hoof. As a forward kick emphasising discomfort at having the girth pulled, it wasn't anything aggressive or forceful but typically, since I hadn't expected to ride, I wasn't wearing my gaiters which would have given me some protection. So, I ended up with a nice big bruise which turned a beautiful purple colour. I suppose that was my karma for being so desperate to get on a horse that afternoon and I made sure Cabeza never came to know of it because I knew that if he did, I'd never have heard the end of it.

When we returned to the *casco* I bumped into Pato, who warned me he had been scolded by Cabeza for letting me ride. He shrugged it off and found it funny, especially when Cabeza then stormed over and told me off in front of Pato, before scolding us together like schoolchildren. I didn't take it personally, which seemed to frustrate him even more. Fuming at having been disobeyed, I left Cabeza to cool off drinking his *mate*. I made eye contact with Pato as we passed and we smirked at one another as young children would when they've done something wrong and know it. On the verge of bursting out laughing, we hurried on away from one another and went to our rooms to sort ourselves out before reconvening at the *matera*.

As I've mentioned a few times, this place was riddled with all sorts of wildlife, to the point that I don't think I've ever seen so much diversity in one place. As the days went by, I would see more and more creatures: red deer, wild boar, water buffalo, black-buck, vipers, fake coral snakes, skunks, *jotes* (black vultures), *caranchos* (caracaras- a type of falcon), maras (a strange animal which looks to me like it's half deer, half rabbit), rheas, otters, mouflon sheep, a vast array of aquatic birds like geese and glossy crimson ducks, flamingos, armadillos, vizcachas, hares and burrowing owls. There was also an abundance of foxes, which looked more like little wolves than foxes due to their silvery fur and long, bushy tails. They were a frequent sight, prowling around leaving tiny paw prints in the sand behind them. In addition to all this fauna, in the *monte* and at the *casco* where there was an abundance of trees, the air was filled with the songs of birds, very different to the sound of an English country garden, but pleasant in its own right, as the mix of sounds was wilder and less domestic. Mostly the sky was filled with the sounds of the *cotorras,* a type of small green parakeet, who flew about gladly making their pleasant screeching noises all over the ranch, as well as being a real nuisance to the windmill maintenance men since they like to build their huge nests made of thorny *caldén* twigs right at the top of the windmills, which can restrict the spin of the mill. There was also another type of bird which I never tracked down and got to see, but it would drive me mad whenever we worked in the *monte* since it made a whistling sound so similar to that of the gauchos that I would often worry I'd strayed away from my path and veered into the one of a fellow co-worker.

Unfortunately, I didn't get to see any *yararás* (a type of pit viper) but I did see the fake corals and vipers. In fact, Pirulo and Cabeza found a recently dead viper in perfect condition so as a very cruel joke, they placed it at the entrance of the *matera* to give anyone who entered the fright of their lives. I also regret not having seen any pumas or the even more elusive maned wolf (or a*guará guazú*, as it's called in Argentina), which so seldom prowls the plains that at one stage it was considered extinct in *La Pampa* until it was sighted again a few years ago. Regardless, given the quantity of creatures I did see, I wasn't too disappointed, although I did really want to see a puma. We came within close proximity of one during a morning out riding when Pavo spotted some fresh tracks in the sand but, it was clearly too far gone to track down. Ricardo, on the other hand, spotted one whilst out deer stalking during the shooting season, which I was very jealous of. However, I did get to see something else that was very special in its own right.

When I arrived I had been told that the ranch was home to a breeding pair of Chaco eagles, or Crowned eagles, as they are sometimes known due to the tuft on the top of their heads. With only around 1,500 individual birds known to exist, I was informed that this endangered bird is perhaps the rarest of all threatened raptors in South America and, as a testimony to their survival, only Ricardo knew where the nest was. He had come across it one day whilst out working and in order to ensure its protection, refused to disclose its location. One day, by absolute chance, a group of us were being driven back to our horses after lunch and I distinctly remember being squashed in the back of the truck (as usual) next to El Tío. He was just gazing out of the window as we drove along until he suddenly exclaimed:

'Look, a Crowned eagle!'

I immediately turned to look at where he was pointing and right there, perched at the very top of a *caldén*, was one of these creatures, totally unmistakable due to its large eponymous tuft revealing its identity. I asked Cabeza to slow down a little so I could get a better look but the eagle turned its majestic head towards us and having noticed it had been spotted, spread its wings, took flight and in a few beats, it was gone. It was too far away to have got a decent photo of, but missing this photo opportunity didn't upset me because when you are gifted with the opportunity to see these wonderful, threatened creatures, the overwhelming joy is enough to keep the memory alive in you forever. Just to think that out of the 1,500 individuals that exist of that species, I got to see one.

I really admired the appreciation the family had for the ranch and how much they enjoyed being there. As many of us will know, unfortunately, there are times when you get so used to somewhere that over time, it can lose its charm and no longer be special to you. Thankfully, for Miguel and Patricia, *San Eduardo's* beauty never faded and they always enjoyed their time there to the maximum. In fact, I got the impression that every time they went, they would find something more to fall in love with. They would go out cycling, running, riding, occasionally shooting and hunting, or spend a few days at the lagoon hut – the one where they had their small motorboat and would enjoy swimming and water skiing. Patricia would often tell me about the wonderful wildlife she would see on her early morning walks and show me photos of sunsets and sunrises. She also said she would often get up at dawn during the rut just to sit and listen to the stags roaring. If she was lucky she would spot one, perfectly hidden amongst the *caldenes*. Patricia also told me she had more than once spent an entire night in the hunting hideouts to see boar and other creatures come out of the undergrowth, just to sit and observe their unfamiliar nocturnal activity.

One of the activities the family particularly enjoyed was spending the day fishing at one of the lagoons. Whilst the family were at the ranch for the lockdown period, they gave me the opportunity to go fishing with them on several occasions and just as with hunting, I had never fished before, so I was keen to learn yet another new thing. One specific lagoon in *Etcheto* had a fruitful population of *pejerrey* – an Argentinian silverside whose name loosely translates to 'king fish', probably due to it being held in such high esteem, particularly in Chile and Peru. This one lagoon was therefore the designated fishing spot, of which there were fewer places so lovely where to spend the day. The lagoon was large and well-hidden between the tall sand dunes, with reeds growing on the banks which provided cover for many different species of birds and, to my absolute pleasure, otters. Despite having never seen an otter in real life, I fell in love with these adorable little animals when I watched 'Ring of Bright Water' as a little girl, at the young age of 6 or 7, so finally getting to see some in the flesh, happily swimming in the clear water and basking under the sun was just yet another gift to me from this remarkable place.

The first time I went to the lagoon, the weekend was warm and still. Quite a few of the workers had stayed at the ranch that weekend, so a fairly large group of us were there together; the four members of the family, Ricardo and his son and then myself with Cabeza, Enano, Pirulo and Ruso. The trucks were stocked full of fishing equipment, snacks and food, and we spent the entire day there from late morning until sunset doing whatever we liked. At first I just sat at the tip of one of the dunes, overlooking the lagoon and read the book about Juan Alberto Harriet that Patricia had lent me from the hotel. It was so quiet; the only sounds came from the happy conversation of my entourage and splashes of water from the birds, otters and oars. There was a little rowing boat that Miguel, Lucas and Ruso had hopped into, but I was most pleased to see a 2-person kayak. I love kayaking so the temptation was too much to resist. After asking around, I managed to find a partner and hit the waters with Ricardo's son, who was the only one keen and brave enough to give it a go. We paddled our way to the other side of the lagoon, where we wandered around the dunes and looked across the water from a different perspective before watching a small herd of deer peacefully grazing in the distance. As soon as they caught wind of us they disappeared in an instance, so we got back into the kayak and almost immediately after pushing the kayak off the bank and into the water, two otters swam up towards us - one on each side of the kayak and in perfect synchronisation, elegantly yet playfully, dived into the water ahead of us, no more than an oar's reach away.

With snacks to pick at, the warm sun beaming down and Ricardo preparing a wonderful *asado* for late lunch over a wood fire near the water's edge, there was not a lot more you could ask for. The peace, the silence and the good company was just paradise. Whilst at *San Eduardo*, I often found myself just gazing out across the landscape, taking it all in and just admiring it, thinking to myself how exceptionally lucky I was to be in such a wonderful place. These thoughts constantly crossed my mind whether I was sat at the lagoon or on a horse out working – *especially* whilst on a horse out working. The perfect tempo of the horse's gait, the creaking of the leather stirrups and the calls of the surrounding wildlife meant the tranquillity and landscape of this hidden treasure was the perfect place to just think, clear your head and appreciate everything around you.

As we helped ourselves to the succulent meat and a delicious array of salads, everyone sat together talking. During the conversation, I asked how frequently pumas are seen and how much of a problem they are to livestock, to which the answer seemed to be that at *San Eduardo*, they aren't common enough to be a big issue but on other ranches, especially those with loads of sheep, they can become a problem.

'Apparently,' boomed Miguel, very matter-of-factually which silenced the rest of us, 'on some ranches they keep donkeys with the sheep because pumas don't like them, so they keep away.'

'Really? That's most bizarre!' I replied, totally surprised at this very strange concept. 'Does it work?'

As I looked around a few of the gauchos and Ricardo were nodding.

'It seems to. At one ranch I worked on they had a few donkeys for that very reason and they never had pumas on the property,' commented Ruso.

'How curious. Then again, I used to have a donkey and I've never had any issues with pumas attacking my sheep,' I added mockingly and a few laughs spread throughout.

'Well, there you go! Must be true!' exclaimed Miguel, evidently thrilled.

To be fair, I don't blame the pumas. My donkey was mean; I wouldn't go near it by choice, so I think the pumas are very wise keeping away from those grumpy animals. His back legs would skim past your face quicker than you could react to him lashing out. Pumas do well to avoid these grumpy asses.

After lunch, Cabeza, Pirulo and Enano decided to call it a day and went back to the *casco*, whilst the rest of us stayed until the very end of the day. Miguel, Ricardo and Rodrigo tried their luck out on the little boat, whereas Ruso wanted to try fishing from the kayak. Having seen how keen I was to go kayaking earlier, he asked if I'd go with him and teach him, which I gladly accepted on the one condition that he wouldn't capsize us or catch me whilst throwing out the fishing line. With a very unconvincing 'I'll try my best not to,' I risked it. The moon and Venus were beginning to glow brightly in the sky and were reflected by the water, like a mirror, which had become the same colour as the setting sky. As we took the kayak out and Ruso cast the line, we then sat quietly and motionless, hoping the fish would bite. The clear water surrounded us, perfectly still like a pearlescent sheet of glass rippled only by the recasting of the fishing line. Amongst the serene tranquillity was the gentle sound of the birds entering the reeds for the night and the cattle lowing in the distance. Unfortunately, Ruso was unsuccessful so we returned to the shore empty-handed, helped pack away all the equipment and returned to the *casco* with Ricardo.

The second visit to the lagoon was more spontaneous and there were far fewer of us. This time, only the family and I went with Ruso and Pato. We all squashed into one car; Patricia and I were in the back seats with Pato, who had Ruso on his lap. Poor Lucas had only his head visible from under all the food and fishing gear piled up on top of him in the boot, whilst Miguel drove with Rodrigo on gate duty. Once at the lagoon, Ruso, Pato and I thought it would be fun if the three of us went in the little rowing boat. Ruso was the first to try rowing and he was an absolute disaster. Pato and I were doubled over laughing as he tried to get us to the centre of the lake with little avail, mainly because he was managing only to get the boat to go sideways or round in circles. Eventually he gave up and handed me the oars, which to my surprise went pretty well. Without too much hassle, we got to the centre of the lagoon. We were there for a good while, the two boys fishing whilst I mainly just sat there chatting to them, barely doing any fishing myself.

Suddenly, Pato abruptly broke off the conversation and began reeling in his line. The further he reeled, the better we could see this great big fish on the end of the hook. Ruso got ready to help catch the fish and bring it onto the boat but just as it was at arm's reach, it escaped. This was immensely frustrating, mainly because Miguel would never believe us. Miguel took fishing days as a competition and with no proof of capture, he would shrug away any excuses that a fish had got away.

Upon being called to eat we returned to shore empty-handed but full of laughs. Whilst we had been on the boat, Ricardo had arrived and had prepared yet another mouth-watering *asado* for us all and given how hungry we were before getting onto the boat, we didn't hesitate to reel in the lines and head back to shore. Pato believed it was his turn to row and, to some degree, he was just as bad as Ruso, but in a different way. Whilst Ruso tried to strand us in reeds the whole time, Pato didn't advance a single inch. Instead, we just went round and round and round and all became dizzy to the point that I was surprised that the little boat hadn't been sucked into the water by the whirlpool Pato had created. Ruso then suggested they take an oar each, which made me fear for my life. I really thought that at this point we were all going to end up in the water somehow but thankfully – somehow - we didn't. In the end, Pato gave up and handed over to Ruso entirely, who was as much of a liability this time as he was at the beginning. As we approached the banks of the lagoon, Pato and I noticed we were steadily drifting towards the reeds.

'Right Ruso, go right. We're going to end up in the reeds.'

'Yes, yes I can see that, I'm trying!' he snapped back in frustration.

A few strokes later, he sent us with some force directly into the reeds where we got tangled up, needing to jab and prod with the oars, attempting to get us back out onto the water. Having had his pride dented, Ruso refused to hand over control in an attempt to redeem himself but finally, despite Pato and I aggravating him and making it harder for him to concentrate, we reached the bank and tucked in to some lunch. Pato and I then went out in the boat together and he happily left me in control of the oars. I lost three big fish, got frustrated and, not being the most patient of people, I consequently gave up whilst Pato went and took out seven big, fat ones. Although I had given up on fishing, I did help Pato catch the fish from the boat to prevent them from getting away. At one stage, Pato stood up, slipped over and fell onto me, causing us both to almost go overboard. We burst into hysterics, so hard that our stomachs hurt and the fish at the end of the line had no trouble releasing itself and getting away. We just continued to laugh, remarking on how funny it would have been for Miguel and the others to have heard the splash if we had fallen in.

Over lunch, we spoke with Patricia and Miguel about the other lagoon they go to with their little waterside hut. As we spoke about water skiing, they noticed that Pato and Ruso seemed very interested in the idea of trying it.

'Well, why don't we go over shortly and you can have a go?' suggested Patricia.

We packed everything up and left the lagoon early to head over to the other one, which was only a short drive away. Unfortunately, the evening changed to become chilly and cloudy and although Pato and I weren't particularly keen on getting into the cold water by this time, it did not dampen Ruso's enthusiasm. He put on a wetsuit and hopped onto the boat with Rodrigo as they skimmed along the surface of the water at quite some speed, Miguel trailing behind at the end of a rope on the skis, testing the water. It was obvious they spent quite a lot of time down there because Miguel expertly followed behind the boat, showing off his skills by standing on just one ski as he passed us on the shore, with Lucas paddling in the shallows without a wetsuit, waiting his turn to have a go. Whilst this took place, Pato and I played a game on the sandy shore but stopped as soon as Ruso's turn came up. We didn't want to miss a single second of it.

He waded over to the boat and, with help from Lucas, he put the skis on. Miguel was now on the boat with Rodrigo and was explaining what the starting position was, along with the starting technique, how to stand up and what to do to move off.

'Just hold on tight and move with the skis. I've seen you ride, just pretend it's a bucking horse and you'll be fine,' said Miguel. 'Ready?'

Pato and I certainly were, watching attentively with anticipation. We had between the two of us placed our bets on how long he'd last upright, both of us perhaps a little on the mean side and more or less of the same conclusion that he would stand up and immediately be pulled forwards face-first into the water.

We were wrong. As the boat moved off, Ruso was pulled slowly out of the water but because he leant too far back, his skis came up out of the water in front of him. As soon as he emerged from of the water, he went straight back under, but backwards. Pato and I laughed hard and then even more when Ruso frantically flailed his way back to the shore, as if being chased by something that wanted to eat him. Very humbly he said thank you to Miguel and Patricia for letting him have a go, but that it was perhaps a bit too cold to go in for a second attempt. I felt bad about laughing quite so much so I thought I would tell him that I admired his bravery, but the image of him falling back into the water and frantically flailing back to shore was too much and the memory of it would set Pato and me off laughing again. The boys dried off, got changed and Miguel took Pato and me for a quick tour of the lake on the boat. Sadly, because it was cloudy, we weren't able to watch the sunset but either way, we stayed a while longer, drinking *mate* and snacking on the delicious *alfajores* Juli made. It got dark and as I sat back enjoying the evening I saw a faint light hovering around, not too far away from us.

'Oh my gosh, is that...' I stammered and pointed at the flying object causing everyone to turn and look.

'Ah, a firefly!' Patricia proudly replied.

In films, fireflies always look so romantic and serene and they were yet another thing I hoped to see at some point in my life. Finally, here I watched it in total fascination as it gently floated about. It may have only been just the one, but I was still in awe.

'You don't see them a lot but as you can see, they are about. Just another one of the wonderful things you can see in *La Pampa*,' remarked Patricia.

The ranch provided the perfect environment for fireflies, given that they usually reside near areas of standing water in warm climates with long grass but despite the thousands of hectares, their existence is threatened. Around the world, firefly populations are hugely at risk, mainly because of the spraying of crops and contamination of water sources, which is tragic.

Despite all these visits to the lagoons throughout my time, I caught very few fish. In fact, I think I only successfully caught one, the day before I was set to leave the ranch. It was my departing gift I suppose and luckily Pirulo, Cabeza, Chango and Pato were all there to see it. I saw activity right at the water's edge and cast the hook literally three metres away from where I was standing. I had Pirulo on one side and Cabeza on the other, who were throwing their lines far out across the water.

'What are you doing? For that you could just skip using the line and catch it by hand,' mocked Cabeza, who quickly changed his tone and tact when I pulled this little fish out of the water from right in front of us all.

Sadly, poor Miguel would seldom be as lucky fishing as these chaps were on the occasion of my last day fishing. Miguel (much like myself) struggled to catch anything that day and so would pick out the biggest catches that others had returned with and take photos with them all. Patricia would then upload the photos to the family group chat, making it appear as if he'd caught all of them. Cabeza, Pirulo and Chango, however, had a bountiful fishing session that day. When I had come off the boat with Pato and Miguel, I wandered over to where these three were to see how they were getting on.

Between them they had over 20 fish in the bucket, sometimes with one at the end of all three hooks when reeling in their lines. The funniest one was Chango, who was using a fishing rod with an old food tin as a reel. Holding the rod and guiding the line with one hand, he would reel in the line by quickly winding it around the old tin with the other hand. This lack of specialised equipment did not cloud his success; he would cast the hooks into the water and, seconds later, he'd be pulling in one or two fish at a time. Whilst winding in one particular catch, we saw that the fish he was bringing in was a particularly big one. To avoid losing it, Cabeza stood on the water's edge ready to grab the fish as it came to the surface. He grabbed the line, successfully pulled the fish out and proudly walked over to Chango to give him a big pat on the shoulder in congratulations. The catch was unhooked and as Cabeza walked over to put the fish in the bucket (just like in the cartoons), the fish shot out of his hands like a bar of soap, slapped him across the face and all too soon, trying to get hold of the fish was like a comic juggling act as it bounced from hand to hand in the air each time Cabeza tried to catch it. Pirulo and I, who were sat on the ground watching, had tears streaming down our cheeks and, having consequently toppled over from the laughter, rolled about cradling our aching stomachs. However, the best part of all that had to be the fact that I happened to be filming and caught the entire thing on camera, which meant Pirulo and I would torment Cabeza by playing the clip over and over again, showing it to everyone in order to humiliate him.

Sergio, his wife and their little one-year-old son came to enjoy the evening at the lagoon with us and along with Ricardo, we all sat around the fire as it got dark, sharing *mate* and eating vizcacha in *escabeche* on bread and delicious venison fillet hams preserved in oil with rosemary, made by Ricardo. We sat together eating, sharing and watching the sun go down as we all huddled around the warmth of the fire, enjoying the peace and tranquillity of the atmosphere on this ranch. It was during these moments that you felt all your troubles and worries cease. In a couple of days, I would be driving to the city with document after document of permissions allowing Miguel to take me into the city and to the airport. However, even under these circumstances of repatriation and global pandemics, at that very moment, nothing else seemed to matter in the world. All that was important was the enjoyment and relaxation in good company surrounded only by wild, unspoilt nature. The darker it became, the more the night sky would become filled with the delicately twinkling illumination of the stars. I had never seen a display of stars quite as impressive as there and in the Outback in Australia. The absolute lack of light pollution (or any pollution, for that matter) in those huge, isolated locations opened my eyes to displays I simply could not have imagined otherwise, to the point that it often looked like there was more white in the sky than black.

As the food ran out and the fire died down, it started getting a bit too cold so we packed up and returned to the *casco*, where the successful day's catch was the only thing on the menu for dinner. In the *matera*, the gauchos had fish frying on the stove and grilling on the *parrilla* and the following day, Juli prepared the leftover fish (for there were many) fried in a herby batter for lunch, meaning nothing of the previous day's catch was left after this.

GAUCHOS IN THE MIST

We had a few long weeks coming up as all the cows on the ranch were due for vaccinations and pregnancy tests, which meant that there was plenty of weaning left to be done in the run-up. It was a rather cold Monday morning and a group of us were riding over to *Las Nutrias* to help Gecko bring some cows and calves in and then help him with the weaning. For that week, I was back on Pirulo's Zaino Panzón as I had run out of suitable horses yet again. After bringing one herd of cattle into the pens, we set out to bring in another and as we were riding on the track along the fence line, Gecko made a comment about how my Zaino Panzón was trotting alongside him instead of walking, his little legs trying to keep up with the long, powerful strides of his young horse.

'I know, he's not the fastest of horses. I can't believe that Pirulo told me he is the fastest horse on the ranch.'

'Him? The fastest horse? Don't make me laugh! That horse Pato is riding will keep up with a rhea for miles, so well that you could practically grab the bird by the neck,' laughed Gecko as he nodded to Come Culo, whom Pato was riding that day.

221

'Don't look at me, I didn't say it. I'm only telling you what Pirulo told me.'

'Bet this foal of mine would thrash that horse of yours,' he said with a grin as he looked down at his stunning dark bay that he was riding once again.

'Are you challenging me to a race?' I asked him, playfully.

'Only if you think you can handle it.'

'Course I can handle it *loco*!' I yelled without hesitation.

As the tension set in, the music from 'The Good, The Bad and The Ugly' may as well have been playing as Gecko adjusted his *boina* and I tightened the strap on my hat. We walked side by side, the horses getting skittish having felt our change in body language, and Gillo counted down for us. At the word 'go', both horses leapt forwards as Gecko and I dug our heels in. The Zaino started off strong and we bounded ahead. As I quickly glanced over my shoulder, I thought we were maintaining a safe distance but in no time, Gecko's youngster had upped his pace and he shot up alongside us. We galloped head to head for a few seconds until the long, powerful strides of Gecko's colt outran the short, quick strides of my poor little Zaino, leaving us in the dust. I pushed my little tubby horse one final time hoping he might make a miraculous bound for victory, but Gecko got away. Admitting defeat, I pulled up the Zaino and declared victory unto Gecko, who was looking very smug indeed. Up ahead he stopped, waiting for me to catch up.

'Hmm,' he grunted as he looked down at his horse's neck. 'That's the first time I've galloped this horse like that. He's pretty quick, don't you think?' he joked as I slipped my reins to let my horse catch his breath whilst we waited for the others to catch up with us.

'Well done... but try me again when I've got a quicker horse,' I provoked him.

That week was followed by a miserable weekend. The weather was dreadful and the temperature had started dropping, too. From riding and working in a t-shirt, I was now wearing a fleece during the day, not to mention it getting down to just below 0°C one morning over at *Las Nutrias* on one of the days we were down there. There were even a couple of nights where I lit my little wood burner, meaning I went to sleep with bright orange cinders gently flickering away in the corner of my room – a sight that would fill anyone with warmth and cosiness. Of course, I hadn't expected to be at the ranch for such a long time, so I wasn't equipped with any warm clothes at all. However, everyone was amazing and I was incredibly well attended to; Patricia had given me a couple of extra fleeces and a thermal underlayer that I could use for work and the gauchos let me borrow jumpers, gloves and coats as well. I felt so well looked after, and I truly was. The family always ensured I was comfortable and happy at all times, Juli forever made sure I was never hungry and that my room was comfortable and warm and the boys offered their coats and jumpers at every opportunity. Needless to say, it was around about this time that I really began appreciating being squashed in the middle seat in the back of the truck, huddled up between the thick coats the others all wore, warming myself by sapping the heat from them like some sort of reptile.

It began raining on Thursday night and carried on until Monday, stopping briefly on Sunday morning. With a busy upcoming week at *Etcheto* and having tired out the Zaino Panzón with that race against Gecko, I needed a new mount so Pato and I decided to get horses from *Don Juan* on Sunday and ride them over, ready for work the following day. He lent me his lovely old chestnut mare, Gerenta, and he grabbed his big, handsome dark bay Gusano - the same one he rode at *La Vigilancia* when he brought over the Alazán Viejo for me.

'She's a real sweetheart. Due to being an old girl, her gait feels stiff but she'll do you just fine.'

Gerenta definitely debunked the stigma around chestnut mares. She was a big, chunky horse, but so genuine and kind that I don't think a single bad bone existed in her body. Having been left unridden for quite some time, not only was she quite fat but her mane had grown long and we didn't have any shears to cut it back with so she was left with it for a couple of days. Out on the ranch, they liked having their horses on the fatter side. Tío Flaco was renowned for having fat horses – all of them were great porky things. Not too fat, because then they would struggle with the work, but not slim because it was amazing how quickly these horses shed their weight when working. Besides, the frequency of work the horses received and the wild life they lived during their holidays meant they never really got unfit. Fat maybe, but never unfit. The gauchos also have this theory that you should always back a young horse when it's on the fat side, because what a horse won't do when it's fat, fit and strong and in training, it'll never do when grown up. On the other hand, if you starve the horse thinking it'll make it easier to train, the day they gain energy and strength, they'll give you a desperately hard time.

As I stood holding Gerenta prior to saddling up, Pato tugged at her mane lovingly with a gentle smile on his face.

'I knew this mare when she was a filly,' he added whilst stroking her rusty brown neck. 'I even watched my father back her. She was always kind and when she was fully trained I then started using her. You wouldn't believe how many times I've taken this horse out hunting and have come back with a boar tied up over her rump. Being quite a big girl she wouldn't bat an eyelid at having one of those huge animals on her,' he chuckled. 'She's about 25 now; we really do go way back.'

He gave her a final pat on the shoulder and turned away to tie up Gusano so we could both get going but typically, the moment we saddled up and set off, the Heavens opened. We tried to wait it out by sheltering under a *caldén* but the rain kept falling, becoming heavier and heavier. The water soaked through my felt hat, drenching my head and despite wearing a waterproof, jacket I was getting very wet. The water off my hat would trickle down onto the collar of my t-shirt. To make it even worse, having put my jacket on whilst mounted meant the bottom of it had got caught behind me on the cushion of my *recado*, resulting in an accumulation of water in a little pocket. I shifted to accommodate my position and my movement released the cold water, which flowed like a torrent down the back of my jeans. Pato looked over at me with a questioning look as I let out a gasp and sat there frozen in position with my eyes wide open. When he asked what was wrong, I rather embarrassedly admitted that a large amount of cold water had just gone down the back of my trousers. As he cackled away, all the little pools of water his coat had collected were subsequently released as well and his *bombachas* also became soaked in the perfect act of karma.

Seeing the sky was not showing any signs of alleviating the rain, we decided to abandon the *caldén* and rode on but being already soaked through with the cold wind blowing there was no way we wanted to get all the way to *Etcheto*. We changed our plans and decided to leave the horses at one of the windmills just under half-way and we rang Petaco to come and find us. Dripping wet and cold, we happily jumped into the truck and completely steamed up the vehicle as we began to warm up. We changed our clothes as soon as we got to the *casco*, where we were met by Cabeza, who wasn't at all pleased with our decision of going out and getting totally soaked through to the skin. Due to the foul weather, the electrics were playing up and when I entered my dark room and hopped into the shower, I was disappointed when I turned the water on and it ran totally cold thanks to the boiler taking the afternoon off. Needless to say, after a very unsatisfactory shower and feeling just as cold as I was when I got in, I changed into dry clothes and went back over to the *matera* to warm myself in front of the fire and drape my dripping clothes over a chair. This also gave me the chance to get to know the newest arrival better, who had arrived a day or two before. His (nick)name was Paisa, and he had been recruited as the *postero* to take over *La Gitana.* He was tall, about 40 and a man of few words but seemed affable and had already started earning brownie points by having prepared a large basket of warm, crunchy *tortas fritas.* After all the pregnancy testing across the ranch was completed, a couple of hundred cows would be taken onto the newly leased property and he, just like Tío Flaco was in charge of *La Vigilancia,* would look after *La Gitana.* This was to alleviate the stocking density on the land at *San Eduardo*, plus having not been grazed for a few years meant *La Gitana* had an abundance of good grass. He had certain elements of character similar to those of Pavo, in that he was laid-back, placid and never gave any trouble. Overall, he was another good fellow.

On Monday morning, Pato and I were driven to where we had left the horses the day before. Luckily it had stopped raining, but it was rather chilly and the whole ranch was covered in a thick layer of fog, so thick at times that you probably couldn't see more than 5 or 10 metres ahead of you. We plodded on regardless and I rather enjoyed taking in the change of scenery and seeing the landscape under such different conditions. Once we got to *Etcheto*, the faint dark shadows of the dunes made what was usually a beautiful, tranquil landscape rather eerie. Along the way, we contacted Pavo by radio, who had arrived at the ranch after his weekend in town and was riding over to help us round up the cattle. We told him our route and current location, and he told us his. Pato and I rode, chatting away of course, until finally we reached our destination, which happened to be the lot with the fishing lagoon. Riding along the banks of the fishing lagoon with the mist hanging over the water's surface made it hard to picture how glorious this place actually was. The days I had spent there were under the nice, warm sun and riding along the water's edge on this grey, opaque day gave the place a totally different, almost spooky character.

We radioed Pavo again, who this time said he was taking a totally different route to what he had originally said.

'But... when we last spoke you were entering 11. How are you now in 8?' enquired Pato, totally confused.

'Yeah no, I went a different way,' replied Pavo casually.
Pato and I began worrying slightly, considering that if he was genuinely in 8 when he had been in 11 before, meant he was somehow heading back in the direction of the *casco*.

'Bet he's got lost in the fog,' he remarked naughtily.
By the time we had reached the far corner of the lot, we didn't have to wait long for the fog to start clearing and the sun to start shining. We decided to wait there for Pavo to arrive, giving the horses a chance to have a breather as well. At 609 hectares, this was a big area for just three horses, particularly with one of them being old and ridden by a not-so-experienced gaucha.

Pato and I waited and waited. His horse had caught my eye the first time he rode him; I thought he was absolutely beautiful and I couldn't resist making the most of the waiting time so I asked him if I could try Gusano out. Knowing I loved trying new horses and that I had a list on my phone with the names of all the horses I had ridden (and aimed to get as many names on the list as possible), he said yes and hopped off.

'Have a canter around here,' he said to me. 'He's quick, strong, responsive and bouncy. He won't take off with you - he's a nice horse, but hold on tight. I don't let just anyone ride him, but just because it's you -' And there he stopped and gave me a quick smile.

I took this as my cue and I trotted off. I only rode this horse for a few minutes but just with that, I fell in love with him. He travelled with his head held high, snorting in his stride with his huge neck arched over and his big, heavy stature combined with his high, powerful shoulder movement truly made me feel like I was riding a horse into battle. We cantered, perhaps even galloped a few strides, and upon returning to Pato I told him that if this horse were mine I'd have called him Bucephalus, since he was the realised image of how I imagine *Alexander the Great's* horse would have been. I sat on him for a while (I was in no hurry to get off) and we continued to wait a little while longer for Pavo, who we still had absolutely no sign of.

'I bet the fog will lift and we'll see him going round and round the same dune,' chuckled Pato.

More and more the fog was lifting, to the point that there was only a fine mist of droplets in the air. We gave Pavo one last call and, deciding he was too far away to continue waiting, Pato and I decided to start. I hopped back over onto Gerenta, was directed to my course and we cantered off. The gate across to the next lot over had been opened a few days beforehand in preparation for this, which allowed the cattle to freely make their way over, so this was just a matter of gathering up the stragglers. About half way across Pato called me on the radio.

'Sofi, come further towards my side, a few cows have escaped backwards between us. Also, we have a Mayday situation.'

'OK, no problem,' I replied. 'A Mayday situation with what, a cow or Pavo?'

'How could you possibly say that... with Pavo of course!'

228

I kicked Gerenta on, laughing as we went. Poor Pavo. Thank Goodness he was such a nice chap, but the fact that he was so nice almost made you feel even worse about the amount of abuse he received. Eventually I crossed the area, successfully sending a few cows through the gate and having been told everything was under control and they didn't need an extra hand, I waited for the other two to arrive. As they arrived, I went over to greet them, say hello to Pavo, and ask him how his weekend was. With a glum look on his face he looked at me out of the corner of his eye and said blandly;

'Why are you two so mean to me,' with a slight smirk on his face.

Pato and I laughed. I put my hand on his shoulder and assured him that it was all just teasing. With quick reconciliation, we headed to the windmill and untacked the horses.

After lunch, Gillo came with us instead of Pavo and all rather suddenly, Pato started feeling very unwell. *Under the weather* may be a more appropriate term considering that, despite his nickname, Pato can't get wet or else he gets ill – something I had no idea about. Thanks to the day before when we had got soaked in the rain taking the horses to *Etcheto* from *Don Juan*, this had taken its toll on him. Regardless, he plodded on listlessly and far more slowly, lacking enthusiasm and using a whistle to herd the cows as his throat was too sore to shout. The three of us moved the cows across yet another lot towards the handling pens where we would be met by Enano and Javier further down to help us finish the move. Gillo, out of nowhere, decided to start picking on me via the radio which always lightened the mood. He would always say something to get a reaction from the others over the radio, like 'Uh oh, Sofi's got lost again,' despite me being just a few metres behind him. Of course, anyone listening in couldn't avoid picking on me, but I would fight back and pick on Gillo in return. With Pato whistling away in the distance, Gillo felt the need to change target.

'What are you doing? Are you umpiring a game of football they're playing over there or are you trying to train them to come to you like dogs?'
As I rode atop of a dune I saw Pato in the distance.

'Doesn't look like it's working. All the cows are running away from him,' I replied on the radio.
Then joined the voice of Javier into the conversation.

'Sorry to interrupt the fun – I'm just going to check around the lagoon for any cows down there.'
Big mistake.

'You watch out at the lagoon Javier,' I said. 'We all remember what happened last time,' and the memory of what happened that day with the evil cow brought a surge of laughter and comments from everyone across the ranch, who now joined in to pick on Javier.

Radio conversations were often funny, given that, of course, anyone could listen and join in. Once, over at *Las Nutrias*, I was by a gateway to guide the cattle through when suddenly, through the radio I heard, 'Quick Sofi, stop the cows passing through the gate!' Thinking it was Gecko, I did exactly that and charged off to change the direction in which they were all heading. Gecko then appeared out of the dust and shouted across to me, saying to let the cows go through the gate. I got confused and later asked him why he had told me to turn the cows around, to which he replied saying he hadn't. It turned out to be Chango, who was listening in on the conversation whilst doing his duties back at the *casco* and thought it'd be funny to interfere. Knowing that I struggled to identify who was talking on the radio, he saw the opportunity and took it. I scolded Chango when I saw him later on, saying all the cattle could have gone AWOL and it would have been my fault, but at the same time I did find it very funny and the fact it would have been all my fault made it even more amusing to him. There was no reasoning with these guys. At first, when Miguel and the family were at the ranch, the boys would lay off bullying me via the radio because Miguel always carried his own and they were probably worried that they would get told off for being mean to me or something. However, as our relationship grew, this holding back became a thing of the past and it would amuse me when I saw Miguel, Patricia and even Sergio and they would say, 'you really do put up with a lot of their rubbish.' Knowing it didn't bother me, they laughed and occasionally there would be a cheer from Miguel, who would be quietly listening, when I came up with a good comeback.

We finished unsaddling the horses just after sunset and we waited to be collected. I was walking with Pato, who by this point was feeling truly rotten, when suddenly I saw a bright green light in the sky. At first it looked like the same sort of green colour that shines on appliances and battery packs, which made me think at first that Rodrigo was flying his drone, but I wondered why he would be flying his drone out there at this time. There was also no noise, given that the propellers on the things make a loud whizzing sound. Gradually, this 'thing' grew larger and brighter until the body of it was about an inch long across the sky – and much longer including the paler green, bubbling trail blazing behind it. I pointed it out to Pato, who only just caught it before it disappeared into thin air a few seconds later. I don't know exactly what it was, whether it was a meteor or comet or what, but it was the most dramatic stellar display I'd ever seen, if indeed I can call it that. It was like something out of a sci-fi film and the size of it, along with its bright electric green colour was simply unbelievable. So unbelievable, in fact, that when I told Gillo, Enano and Javier what I had just seen, they didn't believe me and said I ought to lay off whatever I was taking.

'Honestly! I'm not lying. Pato saw it too, ask him.'

'Yeah yeah, and Pato's about to drop dead so he'll be seeing all sorts right now,' replied Enano.

He wasn't wrong. Pato was looking pretty awful. Luckily, when we got back to the *casco*, Chaque walked into the *matera* and commented on this amazing green light he'd seen in the sky whilst he was over in *Etcheto*. Thanks to Chaque the others now *had* to believe me, but instead, true to form they just teased us both saying that whatever I was on, Chaque must have been enjoying some of it too.

Pato didn't even have dinner that night. After getting in from work he cleaned himself up, went to bed and stayed there. I felt really sorry for him and thought a cup of my magical remedy would help him, but with everyone getting ready to go to bed I wouldn't be able to take it to Pato myself. So, I captured Chaque and got him to help me. I grabbed a mug from the *matera* and went to see Juli, where I asked her for some lemon, honey and thyme. I was lucky, given that she only had a tiny bit left of all these ingredients. I put a kettle of water on the hob to boil whilst I squeezed out the lemon juice, scraped the last of the honey out of the jar and sprinkled some of the dry thyme into the mug before pouring in the hot water. Whenever I've got a cold or sore throat, this has always been my lifesaver and placing all my confidence in it, I handed the mug to Chaque and asked him to take it to Pato with strict instructions: he must drink it all and in the event that he doesn't like it – tell him to man up and get on with it. He must, under no condition, leave any of it, because regardless of how ill he feels I'll give him a telling off before I then send in Juli, who would scold him for wasting the last of her ingredients. Both of these were very real threats.

The following morning, Pato was nowhere to be seen. Chaque came into the *matera* and causally remarked,

'Yup, sick as a dog. He'll be going home to recover, as it's no good having him sitting in his bed here, groaning and festering. It's all your fault Sofi. We all know: *Pato can't get wet.*'

'My fault? Hardly my fault if I didn't know and if that's the case perhaps you should change his name to something more appropriate... like Gremlin or Cat,' I replied.

'No no no,' teased Chaque. 'You made him go out in the rain so it's your fault,' he persisted.

There was no point in arguing. They would continually tease and seldom backed down. Sometimes I could fight back and win, but the majority of the time I couldn't or it wasn't worth it - but not with Chaque. Chaque was particularly good at winning teasing battles. He'd usually just say some stupidity which didn't really have an answerable response other than laughter, a response he would always take as a victory. I may have learnt the hard way, but I did eventually find that sometimes the best answer was to simply nod and agree, or do nothing at all.

Pato's absence meant it was just Enano, Gillo, Javier and me to move these 700 odd cows along the final step of the journey into the handling pens. Claudio, the vet, had arrived to begin the pregnancy testing and had the entire team over at *Las Nutrias* to help him since they had 900 cows down there, which had to be done in just one day. This left us with no extra hands to help us in *Etcheto*. The four of us had to get all these cows into the cattle yards ready for the other vets to come and vaccinate them in the afternoon. Then, tomorrow, Claudio would make his way over to us and test all these cows over a few days. As we were driven over, I thought about this and knew we would have a day with a lot of fast-paced riding in order for the few of us to get all these cows together and I couldn't help but feel that old Gerenta wouldn't be able to keep up. The day before she had been awesome, but she couldn't keep up a gallop for very long and struggled to run up and down the dunes well, which I put down to her rusty old knees perhaps being a bit sore. I was scared that if I pushed the poor old mare too hard, which I felt inevitably I would given the circumstances, I was going to either break or kill her. It would also be very inconvenient if something went wrong, because there weren't many of us to mop up a mistake if the cattle played up. Therefore, when we caught the horses and prepared to saddle up, I discussed my issue with the team and said I thought it would be best if I saddled up Gusano instead. I received a few concerned looks but I told them I had given him a quick try whilst waiting for Pavo and that if I hadn't previously been given permission by Pato himself to ride his most dear horse, I wouldn't even ask. Reluctantly, they saw sense in what I was saying and agreed.

'Be careful though, that horse is half mad and quite nervous,' warned Enano.

I thanked him for the concern but insisted I would be fine and went ahead saddling up Gusano. As for Gerenta, she was left loose to follow us as we went about our business. At first, the sweet old girl followed me wherever Gusano and I went, but as we soon caught up with the cows and the pace quickened, she decided following us was no longer worth it and went about her own business, slowly making her way across with the cattle.

I had Enano and Gillo working to my right and Javier along the fence to my left. I stopped atop a tall dune, watching to make sure the cows I had slowly been collecting were heading in the right direction and I looked out for any more. I then saw Javier appear on a dune far away and somehow, I just made out his words:

'Sofi, come over here!'

Typically, with all the bad weather we had been having, last night had resulted in a spectacular electric storm. It really was quite remarkable and I stood under the walkway outside my room watching and listening to the crackle of the white forks of lightning strike down, whilst also feeling the roll of thunder through my body and the deafening sound of the rain. This meant yet again we had lost all power at the *casco*, so we couldn't charge up our radios. They had all run out of battery from the day before, leaving us today – perhaps one of the days we most needed them – without a way of communicating. Thankfully, despite my rather bad hearing, I made out what had been shouted at me and so I kicked on Gusano. Being my first proper gallop on him, with barely a kick I was taken by surprise at his speed and power. He needed no encouragement to fly so sitting well upright and hanging on tightly, I let Gusano go and he effortlessly covered the ground with such exceptional power that we reached Javier in seconds. This was the fastest, most powerful horse I have ever ridden.

'I'm going to need some help...' I heard him say as I approached. I ascended the dune, pulling up next to him and down below, all packed in the corner, were about 600 cows. 'I've tried pushing them on but there are too many for just me alone,' he said. Looking down at this massive herd of cattle who had blocked themselves right in the corner, I quite agreed with him. Just to add to that day's drama, Javier was riding one of his youngsters, the same little chestnut colt he rode in *Etcheto* the day he went swimming with the cow. With just two of us, one of which aboard an inexperienced horse – I wondered whether we would be able to manage at all. However, since there was no way of calling the others for help without wasting time and precious energy crossing the entire lot to find them, we were left with little choice. We discussed options until we worked out a plan: with the cattle walking along the fence, I would push from behind and Javier would do his best to stop the cows from running out. Two horses, 600 cows and several kilometres between us and the cattle pens, I felt so relieved I that had opted to use Gusano.

With a huge effort, we got the cows out of the corner and moving up along the fence and Gusano was superb. Working this horse really showed me what he was truly capable of, unlike just going for a canter under Pato's watchful eye. He was strong but not ill-mannered and galloped off with exceptional turns of speed, lifting his front end up with such energy that he felt like a rocking horse. When bringing him back to slow down, he would canter on the spot, chin almost touching his chest before reaching a full stop, exactly how some horses do in films when the knight pulls up his excitable steed upon heroically galloping into the castle's courtyard. He was bouncy and fidgety, not what I would immediately call an easy horse to ride due to the power in his neck and his skittish behaviour constantly testing your balance, but a horse's character and strength (as long as not used maleficently with a bad attitude), was something I loved. Full of energy and desire to do his job, this horse was simply incredible.

As Javier galloped up and down alongside the stock, Gusano and I zigzagged behind the cattle, maintaining a constant push. Occasionally, one cow would succeed in her getaway, forcing Javier or me to leave our place in the drive and go after her. I won't deny it – this was seriously hard work. Javier and I were puffing and sweating, not to mention the state the horses were in. However, despite my dark horse being lathered in frothy white sweat with his nostrils flared to the maximum, this beast I was riding did not tire. Finally, perhaps 30 or 40 minutes later or more, we reached the final corner of the lot where we met Enano and Gillo. Between the four of us, we were able to get these cows into the yards without too many problems. There, amongst all the stock, as if she was a cow herself stood Gerenta, chilled as anything, and I am certain that if she had stuck around to see us working, she would have been very glad to not have been a part of it. The stock was shuffled in the pens and the horses were given a very good wash down at the windmill. Luckily for the afternoon session with the vaccinations, Gillo and Enano were going to push the cows on their horses, agreeing that Gusano and Javier's young foal deserved a rest after the work they had just done.

Starting the vaccination of 740 cows at 5 in the afternoon would make anyone think we were going to be there all night but luckily, a few weeks before, we had vaccinated a different group where I was left astounded by having done the 1,074 cows from Lot 13 in just an hour. Yes – one hour. One vet jabbed the cattle whilst his assistant always had another vaccine gun loaded and ready. The crush was left open and the vet would vaccinate the cattle as they ran through. We had two people on horses, two on foot pushing the animals into the race and one person to push on any cows that decided to wimp out and stop inside the race, causing a jam. Considering there were 300 fewer cows than the last time, the work was quite easy and with the cows being tired from the drive in the morning, we finished everything it about 35 minutes. The vets left (giving the human patient a lift into town so he could go home to recover) and it was an early night for us all, given the long days of pregnancy testing we had ahead of us.

HEAD, BODY, TAIL

The morning after our mega roundup it was our turn to test the cows with the vet in *Etcheto.* With the distances out there being so huge, the vet would be offered a bed for the duration of his stay, since driving back and forth from his home to the ranch was simply senseless and unfeasible. Therefore, in the morning, I offered to help Claudio set up his equipment at the pens so we headed over first, followed closely by Enano, Gillo and Javier. Along the way, despite having seen one another when he arrived back to *San Eduardo* the day before, he expressed surprise that I was still there, given that the last time he had been at the ranch we said our goodbyes when he left, since I was meant to have been gone at the end of March. I explained that with the coronavirus and its subsequent lockdown, I was to be there for the foreseeable future, with no talk of imminent flights home.

All the others returned to *Las Nutrias* to help Gecko take all the cows out of the pens and separate them to where they needed to go, except for Ruso who decided to come with us and give us an extra hand, as well as Sergio, who wanted to come and help us out.

These extra hands were very much appreciated, given the number of cows we would be putting through the race, but despite being only six of us and the vet, we managed very well. At first, both Javier and Gillo saddled up to push the cattle into the race, but in the end Gillo stopped mounting up as he was more preoccupied controlling the gate leading into the crush. Javier also neglected the saddle and sat on top of the crush, cutting the tuft of hair on the tail to identify the *cabezas*, *cuerpos* and *colas*. Enano was at the exit of the race, looking into the cows' mouths, checking their teeth to determine their age. As for Ruso, he was at the yoke catching and releasing the cows as well as administering the other vaccines the cows needed (this time, leptospirosis and anthrax). I placed myself at the gate leading into the race and Sergio helped me out with the next gate down. However, with Gillo preoccupied further along and a riderless horse standing around not working, I took it upon myself to do the pushing-in on horseback whilst Sergio did the gates. Although it was more effective and far safer than being in a pen on foot, trapped amongst tens of wild cattle, this was one thing the gauchos were rather cautious of me doing because of the accidents that can happen. They had told me various stories of cattle catching your feet and pushing you off the horse, barging into your legs and breaking them or, in some of the very worst cases, pushing over a horse. Taking all these stories into account, I was very careful and to consolidate this, this little horse was very steady and calm. Besides, being a total beginner to this work I didn't try to work too many cattle at once, so I had far fewer in the small area which reduced my risk of being bashed or hurt.

We had a fire burning to heat up the branding irons, but this also served dual-purpose as a kettle boiler which meant we were able to stop for a breather and a *mate* at around 11, having actually started putting the cows through the race at around 8am. We went back to the *casco* for lunch and a short *siesta* and then returned to the pens. We all worked very hard and despite being only a small team, we worked well together. Everything flowed nicely throughout and having taken it steady, after 300 cows we called it a day as it started getting darker earlier and we still had to make sure all the gates were securely shut and all the cows comfortable for a night in the pens. I know that, as a team, we were much fewer in number but I wondered how on Earth they managed to get 900 done the day before at *Las Nutrias.* It must have been exhausting, especially for poor Claudio.

With many still left to do, we recruited Cabeza, Pavo and Paisa as additional help for the next few days and resumed our places from the day before with a bit of reshuffling: Cabeza took charge of the yoke, this time with Paisa helping out and Pavo did some tail trimming and vaccinating. With his places now occupied by others, Ruso was working here, there and everywhere: this hyper lunatic was running to and from the fire with the branding irons, marking the cows, vaccinating, helping at the yoke, coming over to help me at the gate and pushing some of the cows in. It was a miracle we managed to get through the entire week without him branding any one of us, charging around the way he did with the red-hot irons. At one point, when Ruso had occupied my place at the gate, the vet called me over to him.

'One is so calm and peaceful,' remarked Claudio as he looked up at Gillo, who was calmly, quietly opening and closing his gate whilst gently talking to the cows; 'through you come madam, that's a good girl.'

'Then the other is an absolute nutcase,' he followed up as he looked over at Ruso, who was madly running from one place to another. All you could see was this pink blur zooming all over the place as a result of his bright pink shirt.

'You simply wouldn't believe they're related,' laughed Claudio and I couldn't have agreed more.

All this running around did come to our benefit when Ruso took it upon himself to share out the *mate* whilst we worked. He poured in the hot water and brought it around to each of us one by one, even passing it up to Javier, who today was on the horses. Ruso, on the other hand, was so insane that when we finished and returned the *mate* to him, instead of putting it down he'd be running around doing gates and pushing cows through with it still in his hand. At one stage during his uncontrollable zooming, he came up to me and deposited a large handful of biscuits into my hand.

'I brought these today, seeing how hungry you got yesterday,' he said.

I was hugely grateful. He was right; having breakfast so early and working hard until lunchtime at 1 o'clock meant yesterday I started running out of batteries at around 10, so these little snacks definitely helped keep me going, even though that day we had a change of plan. Instead of returning to the *casco* for lunch, we had brought down some bread rolls and beef which, we would cook over the fire we had made up for the branding irons. It was, of course, Ruso who prepared the meat and put it over the fire. At around 11:30 we stopped and sat down to eat, but the beauty of having lots of meat on the grill meant you could quickly rush over, cut yourself a piece and return to the work you were doing. Therefore, this provided something to munch on throughout the day, which made all the difference to everyone. It really is a natural phenomenon how much better morale is sustained when food is available.

Javier had saddled up both Enano's horse and Gusano in order to alternate, which meant I took advantage of hopping on a horse every so often. At first I rode Enano's 'Rosillo' (strawberry roan - also the horse's name) because Javier had occupied Gusano, but eventually, I managed to argue my way to a horse swap, joking with Javier by telling him that because I'd used Gusano for the cattle drive, technically he was my horse for the week and Javier hadn't asked my permission to use him. It took some convincing, but eventually he gave in and I was able to work this incredible horse in the pens. He was totally fearless and would run up to any cow and barge them into the race with his bold, wide chest and his short, on-the-spot canter gave him that extra momentum to provide greater force behind the push. If a cow escaped us, by spinning on the spot we would turn around and get her back immediately. I had to sit tightly on him to prevent being bounced out of the saddle or slide off sideways because this horse knew what he was doing and would often react to the cows before I did. Directing him with my reins in one hand and my other arm out waving to encourage the cattle meant that all I could do was grip the horse like a clamp with my thighs in an attempt to stay on and move along with him. Meanwhile, this was made far more difficult due to my boots being too thick to fit in Javier's stirrups, so I was left to grip with my thighs alone with no platform.

Thankfully, from all the riding I was doing, my legs were good and strong, so I was fine. Had I done this at the beginning however, it could well have been another story. Interestingly, due to the make-up of the *recados* and how bulky they are, your lower leg barely touches the horse. Instead, during the early days when I was still getting used to this new tack, I found that my thighs would burn after a fast ride, purely due to the way the saddle makes you sit and how it forces you to grip with your upper leg. I was a bit annoyed at Javier because, since he had taken over the use of Gusano, he was tacking him up with his own *recado*, whereas if I had been using my own, I would have been able to use the stirrups properly. On the other hand, Javier's *recado* was so comfy and massively padded that it was beautifully moulded, which kept me in position beautifully, so actually, not having utilisable stirrups was probably not such a problem as I was fixed firmly in place.

I didn't always hop on a horse, though. Sometimes it was just easier if I helped on foot and opened the gates for Javier, rather than both of us being mounted. One time, as soon as I had opened the gate into the race, Javier rode up alongside me on Enano's little horse, put his arm around me and lifted me totally off the ground.

'Come on *Pequeña*, let's go!' he laughed as he pulled me off my feet and cantered off with me helplessly hanging down the side of the horse, much like Gillo that day in *Las Nutrias* when Ruso pulled him off his.

Just like that, he carried me to the next gate where he lowered me back onto the ground.

'Wow, you're strong to have carried me like that' I laughed in surprise.

'You're, umm... Heavier than you look. You must be very heavy boned,' he stuttered, slightly embarrassed which made me laugh quite a bit.

Frustratingly, with just a hundred or so cows left, we were beaten by the sunset, meaning we had to call it a day. Upon arriving at the *casco*, we all went to clean ourselves up and met back at the *matera* before dinner. A few of us were sat playing *truco* when Claudio arrived and said he wanted to join in. He and I played together against Petaco and Ruso and, naturally, thanks to me, we lost quite badly. I found it quite funny and had a good laugh whilst Claudio tried to defend himself, repeating over and over again that he was a good player. Feeling it was best to leave him to redeem himself alone, I got up to help Ricardo and his son sort through four or five baskets full of the enormous pine mushrooms they had been harvesting all afternoon. In turn, I was asked to dine with them and the family, where being a Thursday meant it was *asado* night. Despite having eaten *asado* for lunch at the cattle pens, I didn't mind at all. When the beef was this good, it was a pleasure to eat it all day, every day.

Finally, the next morning we set off for the last day at the pens. As usual, I went over with the doctor (who was still annoyed at me for having made us lose the game of *truco*) ahead of the others but when we arrived at the pens, Claudio's face turned to one of concern. I looked over in the direction he was scowling at to see that the gate between two of the pens had been pushed open and all the cows (who had been separated into *cabeza, cuerpo* and cola) had mixed. This meant they would need to be put through the race *again* just so we could separate them into the correct groups – *again.* Why is it so important to separate the cows? Well, so that when it comes to calving, you know which cows to keep an eye on and when. By knowing the timeframe in which they are due to give birth, you can ensure you have the maximum number of people around to monitor them, whereas if you mix early-stage with late-stage calvers you'll have no idea whom to look out for and can run into no end of problems.

The trouble with running all the cows back through the race was that, as I had said previously, these cows were far from tame: they were harder to work with than the ones you get in England. They made much more fuss going down the race and would frequently, from a standstill, try and jump out over the sides (which were over 2 metres in height). Sometimes, they would be downright impossible to push down into the race in the first place. They would plant themselves and totally refuse to move or circle around the entrance like sharks around their prey. They simply would not budge. Worst of all, most temperamental individuals were always the last to be pushed through as they were clever and knew to stay well back in order to escape being put through the race until the very end. This behaviour was even more infuriating when you had been working for hours and were starting to get exhausted, having little energy left in you to carry on shouting and waving flags at these cows in order to try and get them to do as they were told. It was by no means easy work and to go with that, it could also be quite dangerous.

By the time we had to put the mixed group back through, these cows had gone through the race three times over four consecutive days: first for the vaccine, second for the pregnancy test and the third time to re-draft them. This made them even more unbearable (naturally, as they were getting rather fed up) and even with the horses we were struggling. I wasn't experienced enough so I left the gauchos to do this work on horseback and I helped them on foot, being cautious at all times not to be squashed or kicked by these frustrated beasts.

I was struggling and getting really tired as I tried to push the cows through, as well as jump over the fences to close the gate behind the animals we did manage to get down the race. Ruso, who noticed me faltering, came over and asked me if I'd like to swap places with him and vaccinate instead. I didn't need to think this twice and with huge gratitude, we swapped over immediately. This certainly was a better job, but it also had its challenges. It was incredibly hard to inject the hysterical cows as they thrashed about in the crush, meaning I was being seriously careful not to break the needle or worse, my arm. In farming, a needle breaking off inside an animal is a catastrophe and a member of staff breaking their arm is pretty serious too. If a needle ever snaps off and stays in the animal, the exact location of the needle must be marked on the animal as well as on a record sheet, along with their eartag number and the date it happened. When they go to slaughter, this information must go with the animal so that the location of the needle never goes amiss. Otherwise, just imagine tucking into your steak and finding that. I say that jokingly as it'll never happen, because when a foreign item enters the muscle like that, the body's natural response is to encapsulate it, meaning the muscle around the needle becomes hard, like an abscess, but in the meantime the needle can migrate further around the animal and a lot of meat around the injection site becomes tainted and damaged.

You could of course surgically remove it, but that's rarely even considered. This is all protocol for England, anyway. In Argentina, I'm not entirely sure what they do. Claudio has never experienced it in years and years of practising, which is indeed a very good thing and a credit to him as a vet. Since the cattle are so seldom tagged (because they lose their tags and rip them out so easily), identification must be nigh on impossible but Claudio suggested that you trim the tail hair or tag the animal immediately. Regarding the meat, supposedly the haematomas, bruising and damage to the muscle will alert the butchers and the tainted meat will be discarded. On the whole, to avoid this from happening at all, practically all vaccines are administered subcutaneously rather than into the muscle, which means you are much less likely to get the needle snapping off.

After lunch, we called for further reinforcement once again and Petaco, Pirulo and Chango came to help us, whose fresh manpower in pushing through the cattle came as a Godsend. With their help, we succeeded in getting everything finished and whist Cabeza and Ruso mounted up and went to the open gates ready to let these cows back out to pasture, the rest of us helped the vet pack up. I helped Pirulo, Chango, Pavo and Petaco sort the stock in a few of the pens, moving cows and separating calves. It all went pretty smoothly, but some of these cows were still very angry. One had hurt her leg by charging like an idiot through the mud and because of her pain, she would try to attack anyone who got near her. Others were simply bad-tempered and fed up after so many days in the pens, which I thought was perfectly understandable. Having drafted everyone into where we needed them to go, there was just one cow left in the pen we were sorting and trying to encourage to move, but with so many people around, she got confused. She was about 10 metres away from me with her head down. She looked up at me, snorting, with her ears pricked and a suspicious look on her face. I eyed her up.

'Nah, she won't charge at me,' I concluded, convincing myself that she was only trying to look threatening.

At that moment, she charged at me.

'*Guarda Sofi!*' shouted all the boys at me, whilst they laughed simultaneously. Stupidly, as I turned and ran away, I couldn't help but laugh as well (yes, there's definitely something wrong with me). Luckily, this cow performed more of a 'watch out' charge rather than an aggressive one and she didn't run at me at full speed. After head-butting my arm a few times and *almost* bumping me up onto her neck, I was able to scarper and feeling victorious, she left me alone. We laughed as we could all see she wasn't charging with proper malintent and because I hadn't been hurt, we all laughed even more. The second time she threatened to after me, I wasn't up to risking it. I instantly ran away and jumped the fence without a second thought - an act of cowardice which led to even more creasing laughter from my entourage, so much so that Petaco and Chango, who had just perched on one of the gates, almost fell off.

After battling cows and weaning calves, our day was almost done. When letting the cows out into their designated pasture, we would team up and count them as they ran past, so we could keep a record of how many cows were in each field. This was always like a mini competition and we all had designated places: two or three chaps would slowly push the cows through the gate, only a handful at a time so we wouldn't lose track of the count, and the rest of us would stand in a line counting as the cows ran past. Once the last cow had passed, we would all shout our answer out and if all went well, we would all shout out the same number. If not, usually there was only a difference of one or two, but this would often lead to a heated debate. I remember one particular time when Claudio, Cabeza and I were in discussion over being just one number out. As soon as the last cow ran through Cabeza yelled,

'195 cows – well counted!'

I disputed that I'd only counted 194.

'No no, 195. I'm absolutely sure of it,' insisted Cabeza.

We battled this out between us briefly until we approached Claudio, who said he too had counted 194.

'I would doubt myself but I'm going with what Sofi has said, because she always counts correctly.'

'Fine, fine. I won't argue,' replied Cabeza, obviously frustrated. 'It's only a number, I don't want arguments,' and he huffed and walked away which made the vet and me chuckle. In order to lighten the mood, I decided to tease Cabeza. I put my hand on his shoulder.

'Don't worry, better luck next time. It's OK, I'll help you learn to count,' I joked and in retaliation, he stomped towards me with a faint smile on his face (which he was struggling to mask).

'See how you laugh at me? *Sos mala vos!*' he groaned. I laughed and stepped backwards but our synchronisation was out of time – or perfectly in time, whichever way you'd like to look at it. As I pulled my foot back, my big, heavy boots caught his foot and in the deep sand, he lost his balance and fell over. *Now,* safe to say, our onlookers really were laughing and so was I - until I almost lost my balance and went down with him. Boldly I fumbled and overcame gravity.

I offered him a hand to get up but he refused, got up by himself and put his arm around my shoulders.

'You're so mean to me,' he laughed, then immediately let me go.

As everyone headed towards their transport home I held back, put my hands on my hips and took in a long, deep breath. After several long, hard days at the pens we were finally done and whilst I thoroughly enjoyed every moment of it - I was knackered. As I stood absorbed by my own thoughts and appreciating the silent and empty pens, I felt a hand land on my shoulder. I turned my head and it was Ruso.

'You,' he paused, 'you have worked seriously well these last few days.'

He looked me in the eye as he said this and it's the first time I ever saw him ever express himself quite so sincerely. He was always one to joke around and tease, so witnessing him speaking seriously came as quite a shock to me. It left me speechless until I was able to conjure up the words to say thank you and express how grateful I was for all his help – and the biscuits.

The vet went straight home that evening instead of staying at the ranch, so we said our farewells and looked forward to seeing him in two weeks' time for another round of testing. Back in England, you often hear all sorts of horror stories about ranches in the Americas and Australia not being very nice to their animals, being unjustifiably harsh with the horses, or hitting the cattle in the handling yards. I was pleased to see that this was not the case there. Everyone treated their horses kindly and never hit them unless they deserved a shove in the right direction for behaving badly, which would, as in England, be imparted with one smack, not a repetition of beatings. Some of the chaps did tell me that the odd person had worked on the ranch who treated animals badly, but they were never permitted to stick around for long, which was pleasing to hear. As for the cattle, they worked them in a very different way to what I was used to in England, not only by the use of horses, but with the quantity of animals you were dealing with. You simply could not afford to hang around: the work was much faster-paced and as a result, noisier. It wasn't calm and quiet as is basic practice in the UK because otherwise, you'd never get anything done. When vaccinating animals in England you're talking about 100, maybe 200 an hour if the cattle flow smoothly and you close the gate on each one, whereas out there, we managed almost 1,100. The cattle weren't hit or barged about either, although the odd wild cow would try to jump the fences, not caring whether it was you she landed on or not. In these instances, a piece of blue plastic pipe could be all that was necessary to put her off going any further.

ROASTIES & ANTELOPE SKULLS

Despite the busy week, there was a lot to prepare for over the weekend for the week ahead. So, the hard work continued and as a result, many of us stayed at the ranch over the weekend. We had around 400 calves in the alfalfa to bring into one of the pens ready for the vaccinators who were returning that coming week, as well as separate the males from the females. As with most mornings, in order to get ahead with the sorting, Cabeza had ordered an early start just before first light. Everyone had gone off to get themselves a horse from the *potrero* but I waited with Cabeza because I didn't seem to have one. I sat with him drinking *mate* as we waited to see which horses the others would bring in when suddenly, we both realised Ruso hadn't appeared.

'Go and wake him up, would you?' asked Cabeza.
I turned to look at him with fear. This was, by far, the most treacherous task I had put before me throughout my entire travels. You see, Ruso was good as gold, but his sleep was sacred and he was absolutely not a morning person. He was the stereotypical 'don't make conversation with me right away' and would walk into the

251

matera in the morning, say good morning in as few words as possible and then sit in silence drinking his *mate* and having breakfast. Sometimes, you'd think you waited long enough before talking to him and he'd just look at you, eyes shadowed by his *boina*, and say to you in a low grumble, 'too early'. Sometimes all you got was the look, in which case you knew best to just leave him until he was ready to speak. Therefore, being asked to be the one to wake him up was like asking someone to walk into a lion's den.

'No way. Why should I be the one to?' I replied fiercely. 'He'll bite my head off.'

'After you went to bed last night, a few of us stayed up playing cards and I won 800 *pesos* off him. I am therefore not in his good books and he's more likely to listen to you. Plus, Pavo isn't here to wake him up for us.'

Ruso and Pavo shared a hut and it then dawned on me that poor Pavo had to deal with him and be his alarm every morning. That poor man really was a saint. I wasn't actually afraid of Ruso in the morning. I spent years at boarding school and there were plenty of these grumpy sorts wandering around the halls in the morning and they were the most fun to wind up at breakfast. Therefore, I leapt at the chance of annoying him and, with the grin of a naughty child on my face, I trotted over across the yard to wake him up.

'Tell him he has 15 minutes before he has to be on his horse!' yelled Cabeza after me.

As I approached the hut I began to walk quietly, just in case the commotion of the others preparing their horses nearby had woken him up already. As I crept nearer I could tell it hadn't. I stopped and took a deep breath:

'Morning! It's a beautiful new day and you're late!' I yelled, accompanied by a loud fusillade of knocks.

From within Ruso stirred and amongst grumbling and snarling I could make out the words 'go away,' 'annoying' and 'this early.'

'Come on,' I yelled as I fired away with a few more knocks. We're waiting and Cabeza is going to get angry.'

He grunted and I heard some movement. 'Boil me the kettle would you.'

'The cheek! Only if you appear in two minutes because I'll put it on the fire and not take it off.'

'Fine,' he growled.

Laughing to myself I returned to Cabeza, who was also laughing as he poured himself another *mate*.

'Good job,' and he passed me the *bombilla* as I sat myself down in front of him.

'He asked me to boil a kettle for him, should I - '

'Nope.'

'Cool,' and I took a sip.

The boys had found themselves a horse to ride but seemingly, there wasn't one for me. I told them over and over again that I didn't solely need to ride totally chilled horses but they simply would not have it any other way and any horse I mentioned was a no. Cabeza came over to see who was there and amongst the horses, he spotted 509. His real name, I later found out from a member of the family was 'El Ojito', which means 'The Little Eye'. He came to *San Eduardo* as a polo pony who had passed his peak but was too young to retire from work. Polo ponies are often branded with an identification number and on his left shoulder he had 509, hence that became his name. Quite a few polo ponies came to *San Eduardo* from one of the family's other properties, where they have a few strings of polo ponies. When they no longer worked out on a polo pitch, instead of being wasted, they would head over to the ranch to start a new career as stock horses. 509 had been there quite a while so he knew his job, unlike the horses El Tío had recently taken on, one of which was really struggling to understand her new job, which made poor El Tío's life very difficult. 509 was an old-ish horse, about 15, dark bay and the typical shape for a polo pony, having a slim neck and being overall finely built, but he was much taller than the average horse you'd see on a polo pitch. He was a nice chap, comfortable and quick.

Eventually everyone appeared with their horses ready (having been woken up, Ruso wasn't prepared to speak to anyone until lunchtime) and we got to moving the calves across into Lot 49, which was a plentiful 810 hectares, one end of which was thick scrub, gradually opening up until becoming entirely open at the other end. In the afternoon, we then moved these calves from lot 49 across to the next lot. Between the eight of us we were more or less fine and one by one we broke off into the thick scrub. This was the first time I had worked out on the horses with Paisa and the two brothers, knowing him from before and therefore being allowed to make such jokes, sent him into the scrub and shouted after him, 'Don't get lost!' Thinking I would too, they sent me into the scrub last meaning I had the fence to navigate off but by now I had become pretty good at orienteering my way through these labyrinths and making my way through them in a straight line. Having learnt from the gauchos out hunting, I started using my shadow to navigate, always glancing down to the ground to see exactly in which direction my shadow went before entering the vegetation and keeping an eye on my shadow throughout to ensure I hadn't ventured off course. It was easy to head in the wrong direction when you were having to go around clusters too thick to ride through, or chase a cow that was running off the other way. 509 and I went well and he navigated easily through the trees and bushes but unfortunately, a couple of calves turned around and neither Javier (who was closest to me) nor I could turn them around, which was always infuriating. I, therefore, decided to change tact by pushing the calves out towards the fence instead of having them go through the bushes, which worked... until two clever individuals spotted a section of weak fence and went through it, followed by a couple of others who saw this and decided it was a good idea to follow them. This wouldn't have been so bad if it was an adjoining lot of *San Eduardo* but annoyingly, this was the fence between two properties; *San Eduardo* and a very tiny holding owned by separate people who were landlocked between the sections of *San Eduardo*, *Don Juan* and *La Vigilancia*. Embarrassed, I made the call to Cabeza and we had to return to pick up our calves from these people another time.

It was all an absolute disaster. I was forced to change my mind *again* and pushed the calves away from the fence, making sure they didn't go backwards. Luckily by this stage, we had reached the open area and the calves could sense and see the rest of the herd being brought by the other chaps, which tempted them off in the same direction.

Several hundred calves passed through the gate into the new pasture but everyone had lost some along the way, meaning we were back there the following morning. I saw Paisa and in a teasing way I asked him if he had got lost, to which he replied 'no' and then went off to help Cabeza with a few stray calves. I was stood alongside Gillo when Ruso came up to us and, watching Paisa ride off after Cabeza, he remarked:

'I have no clue what he was doing in there. I saw him cross my path about 50 metres ahead of me and then a few minutes later crossed back the other way.'

Gillo and I laughed and, despite having been blatantly lied to by Paisa, it was very funny.

RIDE LIKE A GAUCHO

The sun was setting so Pirulo was summoned to come and collect us. We left the horses at the windmill, the same place where Pato and I had left the horses the day we got absolutely drenched heading to *Etcheto*. We stood in a line at the water reservoir waiting for everyone to wash down their horses and I waited patiently behind Javier. Enano passed him the container which Javier had filled right up to the top (must have been about 8 litres of water) and he chucked it at his horse. Now I'm not entirely sure how it happened, but the entire quantity of water bounced off the side of his horse and, consequently, I received it all. Javier had broken physics and I honestly have no clue how it happened, especially since I wasn't even stood particularly close to him. Javier's eyes opened wide and he whipped his hands to his mouth as his jaw dropped.

'I'm so sorry! I'm so sorry *amiga*, I don't know how that happened!' he repeated over and over.
With the commotion the others had turned around and saw me standing there, dripping wet and naturally as they always did – burst into hysterics.

'I'm going to kill you,' I said to Javier, looking up slowly through the drops of water dripping off my hat brim. 'I bet you planned that.'

'No, I promise! It truly was an accident!'
As the others laughed, Ruso walked over and put his hand on my shoulder.

'Honestly Javier, how could you do that to the poor girl?' he mocked and then walked off, tutting and shaking his head.
I did feel quite bad because the look on his face truly was one of great guilt. It was very funny anyway and obviously it had been a genuine accident. Of all the people it could have happened to, of course it had to be me - that just about sums up the story of my life, but I can't deny for a second that it wasn't very amusing. Of course I

256

forgave him, but as he handed me the container so I could splash some water over 509 I muttered to him,

'Gosh, what is it with you and water?'

We saddled up the following day to move the calves all the way across the next lot to the cattle pens. This went nice and easily and the gate leading out from Lot 49 was left open for any of the straggler calves to make their own way across. In the evening, we went back and guided around 40 calves which had spotted the open gate and made their way through by themselves and by this point, 509 was already feeling incredibly heavy and tired. We had ridden a fair distance that day and the day before, but not enough for him to be this tired. I worried slightly but upon looking over my shoulder at his rump, I noticed he was definitely on the thin side and, especially being an older horse, was probably the reason he was running out of energy. With all the calves together in the pens they were ready for the following day, where they would be sorted between males and females before the vets were to show up in the afternoon. Cabeza had noticed 509 slacking so told me to let him go in the alfalfa, where this protein-rich crop would give him a quick, much-needed boost. Whilst the others were washing down their horses in the next pen over, I was unsaddling in the presence of two bull calves who were doing everything possible not to go through the gate and join the rest of the herd. Without success trying to get them through on foot, I hopped onto 509 bareback and it was at this point, when haggling with these calves in huge discomfort, that I confirmed – yes, this poor fellow definitely needed some more meat on him.

When everyone arrived on Monday morning, they all tried to find me a suitable horse but there was apparently nothing in the *potrero* that was suitable. Cabeza wasn't going to be doing any horse work this coming week due to other stuff to do, so I asked if I could use his horse.

'No, don't be stubborn. I've already told you that horse is strong and will end up killing you.'

The horse seemed perfectly reasonable and did nothing crazy the last few days, but I could never win an argument with Cabeza. In the end, he asked El Tío to lend me Amarok, the chunky grey mare. The weekend group hopped into the truck with Cabeza and those riding up would bring Amarok to the pens for me.

We started separating the male and female calves, which was eventful and fun. A group from the big pen would be pushed into a smaller pen and from there, we split the males and females in different directions, stopping and guiding them with flags and charging around waving our arms frantically. As had happened many times before, some calves became aggressive when confused and stressed out, which meant a few of us were charged at during this operation (luckily not me this time) and those controlling the gates had to be careful of feisty individuals jumping at them and landing on them. Overall it was a good laugh, but working on the sand was incredibly hard work. The gauchos in their *alpargatas* seemed to glide across the sand pretty easily. I, on the other hand, in my boots, had a bit more trouble. There was a moment when a heifer was running with a group of males towards the male pen and both Ruso and Cabeza were urging me on to chase after her and separate her away. Thank God Cabeza was at the gate and managed to stop them because I put all my power into starting my sprint, only to bury my boot heavily in the sand on my first step, resulting in me tripping over and falling, very nearly flat on my face. Of course this triggered roaring laughter (from myself too) and once he was able to compose himself a bit Ruso, who was nearest me, came over and hoisted me onto my feet, still laughing.

Eventually the rest of the horsemen arrived, bringing even more calves which had been left behind from the roundup at the weekend. This included one individual who had been brought over with a lasso due to his foul temper. Pavo was leading Amarok whilst El Tío led this lassoed calf, which would frequently spin round and charge at El Tío's horse; a skittish, nervous wreck of an animal whose bad temper was being swiftly made worse by this ferocious calf. In that heightened state of anger, there was no way the calf was going into the pens with the rest of them so he was tied up to a post outside, out of the way and left to settle down. In the afternoon when it came to trying to pen him up, he charged at every single one of us, which was quite amusing because of how he'd put his head down and angrily paw the ground before every charge, just like bulls always do in cartoons.

The extra manpower helped a lot but it wasn't quite enough to get all the calves separated before the vets arrived. It wasn't the end of the world, but this meant it took a little longer getting the calves down the race as the steers and heifers required different jabs. Anyhow, we continued working but were beaten by the sunset, so had to call it a day. Starting early the next day, Cabeza ordered only a few of us to stay and put the calves through the race so he sent Gillo, Ruso, El Tío and Enano to round up the cows which were headed to the pens in *Etcheto* ready for Claudio's return the next day. This left Paisa helping the vets at the crush and Pavo and I pushing the calves through.

After putting all the calves from the smaller pen through, Pavo and I went to bring in the rest, which were in the biggest pen. After trying rather hopelessly, there was no way we would get these around with just the two of us on foot. Knowing how chilled she was, I went and grabbed Amarok and with no time to saddle up, I slipped only the bridle on her and hopped on. She did well pushing the calves through and we almost succeeded without any issues until five rebels pushed open a gate into the adjoining *potrero* at the last minute and ran off into it. Pavo blocked the way to prevent any others from escaping and opened the gate for me to get through. The *potreros* weren't huge at all, but they were big enough for cattle to mess you about which meant Amarok and I were chasing them around for a bit before we managed to get them all through. Amarok's wide, chunky back and smooth canter meant I hardly noticed I wasn't sitting in a saddle. She was another great little horse: she wouldn't hurt a fly and, as a bonus, she was small enough for me to be able to jump onto her from the ground, gaucho style. As a result, she was absolutely perfect for me to work with in the pens. What really make me laugh was the faces of the fencers, who never having seen me ride properly before, stopped repairing the fences in the pens and watched me every time I hopped onto my little mare and pushed the calves through.

By the afternoon, we were reunited with the rest of the team yet again and together, we continued back to *Etcheto* to gather yet *another* group of cows to the pens for pregnancy testing. Towards the end of this drive, whilst I was stood near a gateway to help herd the cattle through, Pavo approached me with one of the empty feed bags (the ones they use as a flag to keep the cows moving) and I could just see the tips of two black, twisted horns sticking out of the top.

'Can you take this for me?' he asked.

'No, why do I need to carry your antelope skull?' I enquired.

'It's not mine, it's Cabeza's,' he replied as he helped himself to tying the bag to the back of my *recado*.

'Oh great, so I'll just carry everyone else's stuff,' I muttered. Obviously I didn't care about having the item dumped on me. It was just fun to wind them up, just as they did to me.

'Thank you my dear,' mumbled Pavo, with a silly smile on his face.

At this moment the cows approached the gateway, so we separated into position to guide them through.

Once at the yards, a couple of the gauchos got off their horses to separate the herd equally in the pens so that they all had enough room for the night. I stayed mounted and once all the cows were evenly distributed I saw Pavo who, after closing one of the gates, was trying to navigate his way back without walking through the thick mud. Of course, like all the others, he only wore *alpargatas* and the pens had been churned up by the hundreds of cows into shin deep mud. So, to be nice, I rode over and offered him to hop onto Amarok with me to save him going through the mud.

'Careful, remember I've got your antelope skull.'

'*Si si*, don't worry,' he responded.

I took Amarok alongside the fence so Pavo could mount up from there but she was reluctant to get close enough, so in the end I took my foot out of the stirrup and Pavo mounted from the floor.

'OUCH!' he exclaimed the moment he swung himself onto the horse's rump. I knew exactly what had happened.

'My backside! I just got a horn in my backside!'

As soon as Pavo had mounted up, Amarok started walking off, taking us on her own accord back to the others. Good thing too, because I couldn't see through the tears that filled my eyes and being doubled over laughing meant I wouldn't have been able to navigate anyway. I could feel Pavo fumbling about accommodating himself behind me, the plastic bag rustling as he tried anxiously to remove the skull from the danger zone.

'Don't you dare tell the others what just happened,' pleaded the poor chap.

'That's karma for giving it to me in the first place,' I more or less managed to say amid my fit of laughter.

When we reached the others, Cabeza looked up at me and upon seeing me bright red with tears down my face, naturally, he asked what happened. I composed myself, finally able to breathe.

'No no no – ' I heard from behind me.

'Pavo just sat on the antelope head and impaled his bottom,' which set me off laughing again.

'No! Why would you tell them Sofi?' he uttered in total despair. '*Sos mala vos*,' he huffed and slid off Amarok to get onto his own horse, rubbing his injured quarters as he walked off.

It was all in jest. We both knew that, but almost as if to make sure he wasn't actually annoyed, he glared at me out of the corner of his eye and accompanied this expression with his signature cheeky grin as he walked away.

I had, so far, managed to escape any food prep and cooking, excluding the odd occasion where I would lend a hand to Juli or when I cooked an apple crumble for Miguel and Patricia. Aside from this, I rather conveniently hid away when the boys were preparing food because I simply did not want to be roped in to do the cooking. However, my evasion hadn't gone unnoticed and that weekend, I was pinned down and forced to prepare dinner. To make sure I wouldn't sneak away, they called Juli on the radio and told her not to cook dinner for us. Luckily they gave me a head's up, giving me a couple of days to think about what I might prepare for them, so I spent all weekend wondering what on earth to cook, but nothing particularly special came to mind. What's more, by the time we had made our way back from *Etcheto* and got ourselves cleaned up, it was beginning to get late and time was pressing on and the worst thing you could come across at the ranch were hungry gauchos.

These guys would take 'hangry' to another level, so in order to save myself from their famine-induced wrath, I had to think fast. I ransacked the pantry and Juli's kitchen but I got no inspiration from any of the ingredients until I thought – what if I make something English, something they would have never tried before? They would probably argue against it, seeing as anything English to them meant 'Falklands', but seeing as I was in charge of feeding the troops that night, that was what I was going to do. I thought about making something filling and delicious like a Beef Wellington, but as you might expect, I was missing every single ingredient apart from a beautiful fillet of buffalo that was hanging in the butchery. With no pastry or even enough butter available to make the pastry from scratch and absolutely no mushrooms, sausage meat, or anything similar to use as the filling, a Wellington was out of the question. With a limited selection of ingredients, there was little room for imagination, given that the only thing we had in abundance on this ranch was meat – and potatoes.

As I continued rummaging through the pantry I found sacks upon sacks of potatoes and in a sudden moment of inspiration, I remembered that there was a tub of beef tallow in the fridge over at the *matera*, which had been kept back for making *tortas fritas*. I put the two and two together – roast potatoes! What could be more English? Of course, I couldn't just serve them alone, so I managed to convince the chaps to grab some meat of their choice and cook an *asado* whilst I grovelled, saying I didn't have enough time to make anything proper (and by this point, the Hanger Meter was entering the amber zone).

As I stood in the *matera* chopping up kilos of potatoes, questioning glances were being directed at me from every direction.

'What *is* she doing with so many potatoes? Come on, you can't just make mash.'

I assured them mash was not on the menu, but they were all sceptical. I glanced over at the oven in preparation to heat it up but I didn't have the foggiest idea of how to use it, so I had to call for help. Petaco came to my aid and lit the inside of the oven with a cigarette lighter and told me which dial to use to control the strength of the flame. Talk about outdated appliances. I gazed at the little blue flame and began to get very nervous, as it was pretty obvious that it was going to take a fair while to heat the oven. I looked over at the *asado* and it was coming along beautifully. Would my potatoes make it in time?

After they boiled I bashed them about in the saucepan, as you must, and this attracted yet more attention.

'What on Earth is she doing to those poor potatoes?' I heard Pavo ask.

I could feel four or five of them breathing down my neck as they looked over my shoulder at what I was doing. Eventually, I had to tell them off and order them to go away because it was just question after question. Their absence only lasted so long, until I began lathering the potatoes in the beef tallow. There were so many questions and, by now, *every single gaucho* was stood watching. It was like an episode of '*Ready, Steady, Cook*', but with far more criticism, intrigue and endless questions. It was exhausting. I plunged the tray into the oven and took the chance to run away for a moment because that was yet another thing – you don't use an oven. Pretty much everything they prepared, apart from a green salad, is done on the *parrilla*, so I was being questioned heavily about why I was using an oven. Even after I slipped away, a few of them remained surrounding the oven, still yelling millions of questions in a way which looked as if they were shouting at the appliance, expecting it to answer.

I sat down outside on the porch and El Tío passed me a much-needed can of beer. I took a sip and sat briefly, as I had to keep an eye on my gourmet meal. I couldn't risk it going wrong. Anyhow, finally the potatoes were cooked and most of them were crispy and golden. However, the large majority were only half done because the oven tray I intended to use first was too big for the oven, so I had to take a smaller tray and pile them all up. I grabbed a fork to serve and very proudly stepped outside and yelled, 'food is ready!'

They all rushed in as if they hadn't eaten in weeks, then stood looking totally puzzled at the tray of potatoes.

'What... are these?' enquired Gecko.

'They're called roast potatoes. Very English and typically served with a roast on Sundays,' I explained.

'English? No no no, the English aren't known for their food,' grunted Cabeza. Even the gauchos all the way out there knew that, which is quite embarrassing.

'I don't care. I'm going to try them. I'm starving,' cried Chango as he barged through and helped himself to a good portion. He got everyone going and I insisted everyone try at least one, given all the effort I'd gone through making them. So, like a school dinnerlady, I stood next to the tray and made sure everyone walked away with at least one potato until everyone was served. We then cut what we wanted off the meat on the *parrilla* and sat ourselves down to eat. I went outside onto the porch, where I usually ate, and sat rigidly as everyone took their first bite with intrigue. Yet again, Chango went first and upon watching him take the first bite, everyone else followed. They all looked so serious as they tasted the roasties and the scene was like being surrounded by judges on '*Master Chef*'. Shortly after, I let out a huge sigh of relief as happy 'mmm' sounds echoed around the table and the boys looked around at one another, nodding and chewing away happily.

'These are good!' they shouted in surprise as they munched away. Chango got through his serving in seconds and, before even touching the beef, got up and said, 'I'm just going to get a little bit more.' I was thrilled and the sounds of happy eating could even be heard from inside and from where I was sitting. I watched others get up and go for more, until very quickly there were none left. Not a single fragment of potato remained so I was absolutely over the moon. They must have really liked them because the boys really did comment on how yummy they were without any begrudging, despite the fact that they thought they looked weird and were sure they wouldn't like 'English food'. I was very touched to hear them say all that and I was also very proud of them because it must have been very difficult for them to admit it. Victory to roast potatoes! A true British treasure!

There was no escape now, though. They had managed to get me to cook once so tried to get me to cook again, but I was doing all I could to resist. I was adamant about not becoming the ranch cook. However, they in turn played hard-ball and one day, I had to give in yet again. Some beef fat was being stored in the butchery for rendering and I had some dangerous munchies for some *tortas fritas*, so I decided that I would make them for everybody during the *siesta*. Seeing as I never slept anyway, I thought I may as well make use of my time so before lunch, I grabbed the fat from the cold room and began rendering it all down, saving the *chicharrón* with some bread for us as an apéritif before the main meal. Then, after lunch, I asked El Tío to help me make the dough properly, which he kindly prepared for me and left ready. He then went off to sleep for a bit and I was left with the massive clump of dough he had prepared. I looked around, rather foolishly expecting to find a rolling pin, but had to settle instead for an empty bottle of wine with which to roll the pastry out. With the fat on the boil, I started cutting up the dough into little squares, making the crucial incision in the middle, and lowering them slowly into the bubbling fat. I spent the whole *siesta* time alone in the *matera* just listening to music and getting on with my *tortas* and just when I was almost finished, the boys started coming through. Not all of them knew I was making them, so they were surprised when they walked in to see me stood at the hob, flipping the squares with a basket full of freshly-made golden, crispy warm *tortas* beside me.

'Wow, so all this time we've been asleep and you've been here cooking! Gosh, if I'd have known I wouldn't have gone to sleep,' said Ruso.

'Aw, thank you!' I replied, foolishly taking his comment as a compliment.

'Yes, because I'd have been so scared that you would set fire to everything I wouldn't have been able to sleep anyway,' he laughed and walked away, quickly dipping his hand into the basket of warm treats as he left.

Yup, that sounded more in character. I shouldn't have really expected anything more from him, so being stupid enough to take what he said as a compliment was on me.

As usual, they hovered around me like vultures so each time I finished frying one and put it in the basket, someone picked it out immediately to eat it hot. By the time we set off for work, I think only one or two remained, which were gone by the time we got back. We all knew exactly who to blame. The groundskeeper was the only one who would hang around the *casco* when everyone else had gone, and said groundskeeper always had quite an appetite... so it was obviously Chango.

THE CODDIWOMPLE ENDS

The English language is full of amazing words, but perhaps one of the best ones I know is one I aptly came across during my travels: Coddiwomple. It's a slang word, not found in the Oxford Dictionary, but its definition is said to be 'to travel in a purposeful manner towards a vague destination'. Consequently, this word described my travels well, as I had no idea where I was heading or what I would achieve along the way, but I certainly headed in all directions with strength and purpose. All too suddenly however, it seemed my coddiwomple was set to end.

Upon returning to the *casco* for lunch after a week of pregnancy testing in *Etcheto*, I was met with an email from the British Embassy in *Buenos Aires.* I picked up my phone, hands shaking upon seeing the notification. I opened the email.

'*Good news, two repatriation flights have been organised to return British nationals back to the UK'.*

I remember it vividly. I received the email on Friday and it listed two dates for the flights they'd achieved: the first one being on Thursday, two weeks from when I received the email, and the second flight being the following Tuesday. Further along in the email, they said they would confirm to me on Monday which flight I would be eligible to board - if either of them at all. I should have been relieved to know I would finally be getting home but, in truth, my heart sank and to top it all off, I would have to wait anxiously until after the weekend to know exactly how much longer I had left at this amazing place. I walked to the *matera,* head hanging low and told the boys I had finally been given an indication of when I would be leaving. A few faces dropped whilst others tried to cheer me up by saying that all that was important now was to enjoy my last few weeks to the absolute maximum. Throughout lunch they were great and prioritised speaking even more rubbish than usual to try and get me to laugh and forget about the email. Whilst we ate together it worked, but as everyone went off for their *siesta,* I went to find Miguel to discuss the news.

The events developed at a significant pace. As I had been promised, upon getting back to the homestead for lunch on the Monday after what had been a rather sad weekend dwelling upon the thought of leaving, I had received an email from the Embassy who had placed me on the first flight back. I was gutted, actually. They were prioritising people based on their age, vulnerability and remoteness from *Buenos Aires.* Although I gathered that I was high up on the list of remotely located people, I had rather hoped that there would be a greater number of elderly and vulnerable people which would mean pushing my return date back, albeit only a few days. In the quickest of succession I had gone from having no clue when I was going to be leaving to suddenly having just a week left and all the conflicting emotions left me feeling as if I'd be trampled by an entire herd of cattle. Time had flown by so quickly. I asked myself: Had I really been there almost 3 months? Yet, with the relationship I'd developed with everyone, I felt as if I'd been there for years. I was thrilled to know that after almost 8 months I would be seeing my parents again. However, under the circumstances, I didn't want to go back. Despite the frightening pandemic affecting every country across the world, in this little bubble we all felt safe and were unaffected by the terrible sci-fi-like situation. We were all relaxed, happy, sat next to each other and continued to share *mates*, but since I was going to be on a fully booked plane for 15 hours with more than 200 other people (some potentially infected with COVID-19), I had been discussing with my parents what the procedure would be for me when I got home. It came down to self-isolation for two weeks in my room, no contact with them, no greeting the dogs - nothing. All I would do is spend two weeks, by myself just waiting to see if I developed any symptoms before being allowed back into 'normal' life; or 'new normal', as everyone called it. Until this point the virus was practically imaginary to me, whereas now I began to curse it and the entire pandemic which was causing me to leave liberty and return to confinement, preventing me from sharing all my incredible stories and showing all my amazing photos but, most importantly, keep me from hugging my family.

By midweek, the plane tickets came up for sale. The team were preparing to return to the pens to carry on the pregnancy testing but due to internet problems, I was held up. There were knocks on my door.

'*Sofi*, are you coming or not?' they asked. 'I bet she's sleeping,' I heard them whisper aside to one another.

'No, I'm not sleeping! I have to get this ticket bought. Go ahead without me and if I can join you somehow I will.'

The tickets were sold on a first-come-first-served basis and failure to buy your ticket within a certain number of days meant you would lose your seat on the plane and it would be offered to someone else. As much as I wanted to stay, there was no way I would let this chance slip away and eventually, I succeeded in getting my ticket. Afterwards, I wandered over to the *matera* to see if anyone was around and, for a brief moment, I sat alone until Chaque showed up. Naturally, he asked why I hadn't gone with the others.

'Oh...' He looked down. 'So it's official then. You have the ticket and you're going home.'

'Yup... Monday after next,' I replied solemnly.

For the next few minutes we sat together, not saying much, just sharing *mate* before I asked him if he'd mind giving me a lift to the pens so I could continue helping everyone with the vet. Of course he said he didn't mind at all, so we climbed into his truck and headed over to *Etcheto*. I was quiet on this journey; we both were. Even Chaque was struggling to say silly things to keep the mood light, which was a really bad sign. He usually had no trouble saying stupid things to make me laugh, but he was struggling to come up with anything at all. The poor chap did try, but it wasn't of much use.

He glanced over at me. 'We're going to miss you *Loca*.'

'And you have no idea how much I'm going to miss you, but I promise I will come back,' I replied through a forced smiled.

We arrived at the pens where I got out and thanked him for the lift.

'Never a problem,' he replied. 'Now go on, enjoy and try not to think about it,' and with that I managed to break a smile, hopped over the fences and returned to the others.

They had very few cows left and in a short time, we were done. As always, we needed to count the cows in each group and as we moved all the cows into one pen, I saw a tiny wet calf wobbling, struggling in the mud to keep its balance whilst trying its best to keep up with its mother. I remembered Claudio saying that morning that there was a cow on the brink of calving, so this must have been her calf. The pregnancy test would have induced the labour and whilst we were at lunch, she had calved. To stop the baby from being trampled by the cows, Ruso guided it away from the herd, but it persisted in trying to get back to its mother. Worried for it, I walked up to it and with one arm around its chest and the other under its bottom, I picked the little thing up to prevent it from being hurt and stood out of the way with the tiny creature in my arms.

'*Sos campera vos!*' chanted my friends.

'See? You can't leave now, you've got an infant in your care,' added Gillo.

Going back and helping in the pens was the best thing I could have done, not only because I enjoyed it so much that it made me feel better, but also because you always had to be on your toes, which left me no time to feel sad and miserable. If you didn't pay attention you'd be shouted at for letting a calf slip past when you were supposed to be separating them off, or you risked getting charged at by an angry cow. It had started getting late by the time we had separated the calves off, counted the cows and let them out to their new pasture, so we left it until the following day to load up the weaned calves into the trailer to take back to the feedlot. El Tío, Enano, Pato, Pavo and Gillo saddled up and went to move some cows across onto the adjoining lot, leaving Cabeza, Ruso and me to help Petaco load up the calves once he arrived with the tractor and trailer. He was also going to load up our three horses with the calves to save us the long ride back, seeing as we had come down in the truck. It took 20 minutes to reach the *Etcheto* pens by car, so you can imagine how much longer it took for the tractor to plod its way over. We waited for some time, mostly just chatting and once Petaco arrived, it only took us about 15 minutes before the trailer was loaded with the calves and the horses and he was on his way.

I'm pleased to say that, despite only having a few days left, the experiences and dramas were in no rush to end. Before leaving, we scouted the area to make sure all gates that had to be shut were shut, that we hadn't left anything behind from the previous day and that everything was in order for the next vet session. Then, just as we got into the truck ready to head back to the *casco,* Cabeza was called on the radio by Petaco.

'Have you passed a wheel?' he asked.

'What are you talking about?' asked Cabeza. 'We haven't left yet anyway, we were just about to.'

'Oh, well I'm in a bit of a situation,' stuttered Petaco. 'The tractor has got stuck in some soft sand and in trying to get it out a wheel flew off the trailer.'

'Right...' grunted Cabeza, scowling. 'We'll come and meet you.'

Cabeza and Ruso were ranting about how he could have possibly managed to get himself stuck in such a way, but I just sat in the back and found it all very amusing. As we approached the tractor, Ruso spotted something black in the grass, far off the track. Hoping it was the rogue wheel, Cabeza turned off and drove towards it, where we saw it was indeed just that. The two of them lifted it into the back of the truck and whilst looking around and seeing the tractor a good two or three hundred metres from where we had picked up the wheel, Cabeza got into an even greater fit of rage. This, in turn, was made no better when we got to the tractor and found its back wheels pretty well buried in the sand and the trailer's axle hub where the wheel had come off was totally invisible. Inside the trailer, although the calves looked a bit confused, the three horses stood quietly, making no fuss whatsoever.

'Guess I'll have to go back and get the other tractor to pull all this out,' muttered Petaco. Luckily, the ranch had another tractor (an enormous old John Deere 9020 double-wheeler, in case there are any fellow tractor nerds reading this) to do all the alfalfa cultivations.

'Yes, but we can't just leave the trailer loaded with the stock,' said Cabeza rightly, so he pulled out his radio and called for the others.

It was too dangerous getting into the trailer with it unstable and fully loaded, so the four of us climbed onto it and, through the railings, slipped the bridles onto the horses with quite some precision, skill and teamwork. Together we managed to get the three horses out without any calves escaping. Ruso and I saddled up (luckily we had our *recados* in the back of the truck) and Cabeza drove Petaco back to the *casco* once the others had arrived to help us with the kerfuffle, to unload the calves and herd them to the feedlot, which was a long way away. It didn't take the rest of the group long to join us and with everyone's help, the calves eventually, but with a lot of encouragement, hesitantly jumped out of the awkward trailer and we were on our way. Of course, since Cabeza was driving there was no one to ride his horse, so I offered to lead it back for him.

Along the way, it seemed I had been joined by Indiana Jones. El Tío had brought along a *rebenque* with an extra-long end, just like a bull-whip and was properly cracking it, just for fun. El Tío, Pavo, Pato and Gillo were riding together in a group, where El Tío rode along on the outside, showing off with the whip.

'Let me have a go!' yelled Pavo and it made me laugh when El Tío passed him the *rebenque* and the group immediately disbanded in fear of them or their horses being whipped. The poor fellow tried and tried, but clearly there's a greater knack to it than anyone would think. Plus, he was being very careful about whipping anything, so that probably didn't help his technique. It would have been fun to try but I felt that, knowing me, I would catch Amarok for sure. Plus, Cabeza's horse was a bit of a nervous twit and probably

274

would have pulled me off sideways if I cracked the whip anywhere near him.

As we were almost at the feedlot, I was riding alongside El Tío and he told me a few stories of some cattle drives he helped his family do when he was younger. He told me about one in particular, which was three days long. He, his father, his grandfather and a few additional hands moved around 60 cows from one ranch to another. They asked to keep their cattle in the pens of ranches along the way and, at night, they used their *recados* as pillows and all the saddle rugs, sponges and skins to lie on and cover themselves with, as well as having a couple of pack horses with some food. Similar to Juan Alberto Harriet's epic journey, this too sounded utterly incredible, like the sort of story you would read in a book or watch in an old film but here it was, straight out of the horse's mouth, told by the man who was actually there.

Our conversation was unfortunately cut short when we arrived to the feedlot and our calves began getting excited by the lowing of the other calves already in the pens. They started scarpering, sending us in all sorts of directions in an attempt to keep them together. One bolted in my direction and I could see he was headed towards the gate that led to the *casco*. Without really thinking I kicked on Amarok, forgetting I was leading Cabeza's horse who, despite being all fiery and nervous when ridden, was one of the dopiest, slowest horses to lead. Amarok leaped forwards after the calf whilst the roan decided he wasn't quite so keen and with a massive tug as the slack of the lead rope tightened, I was pulled off backwards out of the saddle. Fortunately, Amarok stopped as soon as I accidentally yanked at her mouth in the process, which allowed me to pull myself off her rump and back into the saddle. I turned around, scolded the unladen pack mule and tugged at his headcollar, hoping he'd pick up the pace. However, all he did was remain calmly at a walk, stretching his head and neck out as far as he could. I of course missed my chance, but luckily Enano was on it and he managed to stop the calf from passing the gate. He was actually right behind me when all of this happened and politely, trying to conceal it, I could hear his faint little giggle when I was jerked out of the saddle.

Eventually, without much help from me because of my dopey lead horse, the calves were integrated into the feedlot and all was well; all apart from one calf, which had decided to do a runner way back and was consequently chased, lassoed and tied to one of the windmills for us to pick up in the back of the truck later due to a lack of cooperation on her part. After lunch, I went with Pavo and Ruso to pick her up, where we securely and firmly tied her to the bar in the tray of the truck and released her with all the other calves.

FINAL ANTICS

This last week went by as a blur. I enjoyed every moment to the max but couldn't help looking around with sadness wherever I went. Would it be the last time I'd be taking in this landscape? I remember we finished the pregnancy testing for the time being and returned to the *casco* with the horses, where we would be swapping them for new ones for the following week. On Friday afternoon, before everyone left for town, I went with Gillo, Ruso and Pavo to look for fresh horses in the hotel reserve lots, which together made up a whopping 1080 hectares. This huge area was the preferred place to release the horses due to the rich, plentiful pasture that grew there. In addition, although it was big, these enclosures surrounded the *casco*, which made it nice and easy to bring the horses in. The three gauchos each found themselves a horse to swap to, but all the horses deemed adequate for me had remained undiscovered, which left me once again without a mount. Since I wasn't in need of a horse over the weekend, I waited until the week to find another.

I asked El Tío where he would like Amarok released so that he would know where to look for her when he next needed her. He asked me to leave her with Cabeza's horses behind the feedlot so after washing her down, I placed my *recado's* cushion on her back (so I wouldn't get all wet and teased by everyone afterwards) and rather lazily hopped aboard. 'Why walk when you can get a lift,' I always think to myself. Besides, this comfy little horse was such a sweetheart that I simply couldn't resist, so with my reins loose and legs hanging down, gently I rode across the *casco,* past the *matera,* through the feedlot and hopped off. I thanked her for her hard work, scratched her forehead, slipped off the bridle and whilst saying goodbye, I released her in the enclosure where she whinnied to the others and trotted off, disappearing into the *caldenes* in search of the herd.

We had a nice quiet weekend doing only what was absolutely necessary, like setting up an electric fence and turning on the water pumps at *La Vigilancia*. Then on Monday, we were greeted with a grey, wet day, which meant we did odds and sods around the *casco* instead of going out in the rain. This just consisted of tidying up areas and fixing things that needed some attention. I took the opportunity to fix my gaiters as the stitching holding the inner padding to the chap itself had come undone and since I had borrowed these from the ranch, I wanted to make sure I handed them back in the same condition in which I had borrowed them. I asked around for a leather needle and thread and eventually, Ruso was able to supply some. I'd never stitched leather before and the boys teased me about how long it was taking me, but I ignored them as usual and carried on at my own pace, finishing the job, which I was rather pleased with. It looked tidy and properly done and the stitching held strong for that final week, which was all I needed.

Whilst doing a thorough cleaning out, Cabeza asked us to take some old bags of wool out of the saddlery and tip them in the pit with the loader of the tractor. I was with El Tío at the time and when I heard this, I jumped at the opportunity and offered to get the tractor.

'Are you sure you know how to drive these things?' they asked with concern.

'How many times – yes!' I replied, irritated to be having this conversation again.

As I walked across the courtyard and into the machinery shed, naturally I was asked what I was doing by those sat at the *matera*. I said nothing - just grinned at them and carried on walking. Chango, who was leaving the *matera* heading in the same direction, followed me and saw me open up the door of the tractor.

'Oh God, the girl's taking the tractor!' he shouted across to the others, who all shouted back:

'No! That's it, say bye to the tractor, boys,' I heard amongst a roar of laughter.

The funniest by far was El Tío. He insisted on coming with me to make sure I knew where I was dumping the old wool, despite reassuring him time and time again that I knew where I was going. Regardless, he jumped into the tractor with me. For no reason whatsoever he looked absolutely terrified as I drove us and his expression became even more hilarious when I drove the tractor towards the edge of the tip where in front of us lay the massive hole the rubbish had to go into. I reckon he must have held his breath as I tipped the bucket, backed the tractor away and took us back towards the *casco*.

'See? Don't know what all that fuss was about,' I mocked him.

With a sigh of relief his expression changed and he gave me his usual naughty grin.

'You're totally bonkers,' he muttered quietly with a smile on his face as he shook his head.

Thanks to the rain, I wasn't in a hurry to find a horse as we weren't going to be out riding that day, so it meant I had time to think about which horses might be available and whom I could pester to lend me one.

'I saw Pirulo's dark bay at *Don Juan* yesterday. He's looking really well, put on some weight,' commented Chaque.

'Por Si Acaso?' I asked.

'I suppose you could have him back if Pirulo agrees,' replied Pato. 'I'm going to go down there to get my buckskin, so you could come with me and get him,'

'Will he be properly rested by now?' I couldn't remember how long he'd be off work for so I looked back at my photos and saw the date of when I'd swapped from him to the Alazán Viejo. 'Gosh, that was 3 weeks ago!' I remarked in surprise.

'That'll definitely be fine,' said Pato. We can take Gusano and Gerenta back and swap them over tomorrow morning.'

At that moment Pirulo, followed by Enano, entered the *matera*.

'Pirulo...'

'Yes...'

'Can I borrow Por Si Acaso again, please?'

Just as I had anticipated he happily said yes.

'I'll come too, I've got one of my *rosillos* down there that I'll swap to,' added Enano.

With that, the following day, Pato, Enano and I rode to *Don Juan* to swap over our horses, whilst the others headed straight to *Etcheto* to round up the next group of cows for pregnancy testing.

Pato called in the horses and just like the time before, the matriarch mare came with the bell around her neck ringing and the rest of the herd following behind her. I was thrilled to see Por Si Acaso again, who cantered in proudly, looking well-fed and rested, just as Chaque had said. He was also much fluffier as his winter coat had started to come through. As always, he stood there patiently, never needing to be chased and once I'd taken my *recado* off Gerenta and washed her off, I walked over to Por Si Acaso, slipped the bridle off Gerenta and straight onto him.

'And there I was, so upset thinking I wasn't going to be here long enough to ride you again,' I whispered to him as I stroked his beautiful, dark face.

The three of us rode with fresh energy on our new mounts to meet the others who had, by this point, almost reached *Etcheto.* We agreed to meet together at the pens of Lot 34 (where I'd done the work with the calves bareback on Amarok) as it was a midway point for both us heading over from *Don Juan* and the others who were approaching from the *casco.* Enano was called to help Paisa on the radio, leaving Pato and I alone at the pens. As we stood there on our horses, I looked over his buckskin and, noticing what I was looking at, Pato dismounted. He knew exactly what was going through my mind.

'Go on then, hop on,' he said with a gentle smile.
With a cheesy grin I hopped off Por Si Acaso and left him to stand quietly as I got onto Pato's horse. He had ridden him for a couple of days at *La Vigilancia* a while ago, but I never really paid attention to him.

The gauchos loved absolutely anything that had to do with horses, especially films about them. I loved this about them and over the months I spent there we had had plenty of conversations revolving around *Black Beauty, Secretariat, Hidalgo* – but the real favourite amongst us all was actually one of my all-time favourite films; *'Spirit: Stallion of the Cimarron'*. As a result of his colour, this cartoon Mustang consequently lent his name to Pato's horse. Spirit was similar in stature to Gusano and, actually, similar in riding style too, but much less bouncy and skittish. He too had a strong, chunky neck, but was noticeably less powerful than his dark bay counterpart. I only had a quick play on him around the pens until the others arrived. Just from this short ride, he seemed pretty cool and energetic. In my opinion, Pato seemed to have managed to bag himself a selection of some of the best horses on the ranch.

Swapping back over to my own steed, together we all rode to *Etcheto* and did what we needed to do over there. One group of cows was to be put into the pens ready for Claudio's return the next day and then two other herds were to be brought closer to the pens, also in preparation for PD testing. That afternoon, as we were rounding up one of the herds, Paisa, Gillo, Enano and I were riding with some cows on one side of a small lagoon whilst Pato, Ruso, El Tío, Pavo and Javier were riding on the other. I was right on the water's edge when suddenly Gillo, who was just ahead of me, shouted.

'Chancho!'

In a second, he hurtled after a small group of wild boar, with Enano following closely behind. The day before, the family had returned to the ranch (as they would be taking me back to the city) and Miguel had said that if we saw any boar, we should catch one so we could have it for dinner. These five boar had appeared out of the reeds and were escaping. One, however, failed to get away and continued to be chased. It made its way around the edge of the lagoon with Gillo hot on its heels, though you'd be surprised how quickly these wild pigs run when you see them alongside a galloping horse.

The boar was now on the other side of the lagoon, where Pato and Ruso joined in on the chase and the boar wove its way between their horses. Out of frustration, Ruso kicked his mare on to be exactly alongside the boar and once head to head, he threw himself off his horse in a very agile manner with incredible speed. Without the horse slowing down even slightly, he proceeded to chase the boar on foot. Now riderless, Ruso's mare slowed down, allowed Pato to overtake and consequently went off minding her own business, clearing off in the opposite direction. Pavo spotted her and was peeled away from the excitement as he chased after the runaway to catch her and bring her back. Much to my utter astonishment, Ruso kept up with the beast for a while. He was seriously fast but inevitably, the boar gained pace and Ruso could no longer keep up. I'm not sure why he decided to hunt the beast on foot... It must have just been part of the fun. Backup then reappeared as Pato charged past, overtook Ruso and managed to get Spirit alongside the boar. With that, he too threw himself out of the saddle, but this time actually onto the boar. He successfully held it down until the others arrived to help tie its legs together. Gillo then called on the radio for someone with a vehicle nearby to come and collect the freshly caught dinner. Paisa and I continued with the cows until they were safely escorted through the gate and onto the other side before joining the others, who were laughing and panting. A few moments later, Pavo arrived late to the party, looking very grumpy leading Ruso's mare. As he handed over the reins he proceeded to tell Ruso off, as best as he could anyway, because no one ever took his anger seriously.

I simply couldn't believe what I had just witnessed. The total madness and turns of events I was experiencing out there just couldn't be made up.

'I'll never meet crazier people than you lot in my life.'

We did a full day with Claudio but luckily there weren't a huge number of cows, so we got them all done in one day. We started early and finished just before dusk, where upon finishing we rode the cows out to their new pasture. It was on this evening that I said goodbye to Claudio again, except this time knowing for sure I wasn't going to see him again. It was dreadfully sad as it was yet another knock on reality, reminding me that my time there was coming to an end. My time glass was getting down to its last few grains of golden *Pampa* sand. In a week's time, Claudio would be back to continue testing the cows and I would no longer be there.

It was Thursday and I was feeling really depressed as my final full week was all too soon drawing to a close. The group of riders split up, sending a few of the chaps to work in one lot and leaving just four of us to work in another. We did our thing, riding around the gorgeous sand dunes, herding the cattle through to their new pasture. On my route, I was met by a *caldén* thicket, too low and thick to ride through. I jumped off Por Si Acaso and knowing how good this horse was, I just left him standing there without even wrapping a rein around one of the trees. After a good rummage through the thorns I inspected the area and with no cows lurking in the shadows, I concluded my search and returned to my boy, who was stood exactly where I left him, his fluffy dark ears pricked forwards as he watched me appear out from the bushes. I remounted and we carried on, this time with the silence of this place absolutely killing me. Usually I enjoyed the sun on my face and the peace of the wilderness, but that day it felt incredibly empty. I had my phone on me, so put on some music in attempt to fill the void. I pressed shuffle and the first song to come on was one of my favourites - *'Iris'* by The Goo Goo Dolls. Now, I will not apologise for

284

the sentiment or cheesiness of this next bit, but honestly, it was almost as if this song was meant to come on.

It's a song I love, yet for the very first time, I listened closely to the lyrics. The opening verse:

> *'You're the closest to heaven that I'll ever be,*
> *And I don't want to go home right now.'*

I wanted to go home, of course I did, but *not right now.* There was so much more I had left to learn from this place and the thought of going back amidst a pandemic was all too surreal to properly process. I'd return to England, not being able to even hug my parents when I arrived. Locked up in isolation for two weeks and then? Who knew. I hoped to find a job but, as we all know too well, this virus had literally stopped the world. I had friends and family who were made redundant, friends who were unable to return to university to finish their final term... I was overflowing with stories I wanted to tell, yet upon arriving, I'd have no one to tell them to. What I'd learnt and experienced from Argentina and Australia was far too much to describe over a phone call and because of this wretched disease the idea of telling my friends and family all these anecdotes over a pint in the sun at a nice pub was destroyed. With no resolution or end to the outbreak in sight, who knew when I would be able to do all this or when life would return to normal.

> *'When everything feels like the movies,*
> *Yeah you bleed just to know you're alive.'*

As these lyrics were sung, I felt a slight stinging sensation on my hand. I looked down to see a small cut, bleeding. It must have happened when I had gone into the trees. I just hadn't noticed. It

285

was true, though: the perfect landscape, amazing people, the tradition, a dream job and the entire paradise element... What was to stop me from suddenly waking up in my bed at home, having just dreamt all of it? There were days so perfect I genuinely feared this. Luckily, I have sufficient evidence to prove to me it wasn't all a dream because, even now, there are times I still doubt it, but I know I am the luckiest woman on Earth because I genuinely *have* been there, experienced it and lived it.

Fighting back tears, I approached the corner of the lot and switched off my music as I saw the other three approaching. They sensed something wasn't right but instead of commenting on it, they behaved as if nothing was wrong and managed to get a smile out of me. They knew I was suffering with the thought of leaving, so they tried to stay well away from the topic. In retrospect, I do think they were doing it for themselves a bit, too.

Friday. Although I still had the weekend, this was to be my last day working on horseback. Much to my relief we worked in *Etcheto* which, almost as if it knew I would soon be gone, gave me the most incredible display of wildlife yet. First we went to work in a lot I hadn't yet been in. As I rode up a tall dune, I stood at the top, gazing out across the new view. Despite being all part of the same property, it amazed me how different the terrain was, sometimes varying hugely from the lot just next door. This one was totally different as it was very hilly yet from where I stood, a large expanse of flat plain lay like a valley between practically symmetrical sand dunes on either side. Beneath me, this vast flat plain was the largest flat area I had seen between dunes in *Etcheto*, and it was marvellous. With no cattle in sight and the boys way ahead in the distance, I pulled myself together and descended onto the plain where Por Si Acaso carried me, charging effortlessly across it, where for a good few minutes all I could hear was the sound of galloping hooves through the whistle of the wind. Once I caught up with everyone, we then moved on to yet another area I hadn't delved into, with much greater dunes and various small lagoons throughout. As I rode on quietly, a large group (perhaps 15 or 20) of bright pink flamingos flew past overhead, squawking as they passed by, their massive wings beating rhythmically. I gazed open-mouthed at this wonderful sight until I was distracted by a shout in the distance. Pato, who was riding nearby to my left, pointed out past me to Ruso, who was approaching us from behind, chasing a herd of deer. As they got closer Ruso slowed down and I sped up. Por Si Acaso flew alongside them flat out. For a while we galloped with them as if a part of them, with a couple almost close enough to touch. As we ran with the herd I counted them: nine does and three stags, each of the males with an incredible set of antlers. I eventually pulled up my horse and watched the deer disappear into the distance as we both stood panting - him from the race and me from exasperation.

All this show of wildlife went on throughout the evening: I saw various armadillos scuttle across the sand to retreat to their burrows and groups of big, beautiful black-buck would appear, bounding across the dunes so majestically that it seemed they barely touched the ground with each leap. It really felt like – and I truly believe it was, as if *La Pampa* in all of her glory and natural magic was showing off her beauty as a way of saying goodbye. We worked on into the evening, not into the darkness, but well into the sunset. The evening air was cool and the sky was at its beautiful orange and pinky-mauve stage as we headed towards one of the *potreros* with the windmills to leave the horses for the night. The boys were in their usual mischievous moods. As we were riding, they started pulling one another's *boinas* off for fun. Gillo pulled that of Pato's off and, as Pato went to wrestle it off him, in Gillo launched it like a Frisbee self-defence. It flew past me and, in reflex, I grabbed it from the air and held it in my hand. This was a mistake. Gillo and Pato both stopped to look at me and upon making eye contact with Pato, his eyes glazed over with a playful evilness.

'Uh oh.'

Immediately Pato started towards me. The expression of sudden regret on my face had everyone howling with laughter but playing along with the game, I galloped off, waving the *boina* like a flag as Pato galloped after me. As I turned to look over my shoulder, the sight of him charging behind me with a look of sheer determination on his face was too much and I burst out laughing, which forced me to pull up Por Si Acaso and resign the hat back to its owner - both of us unable to speak through the panting and laughing.

'Ah, you see? Too scared to gallop off at full pelt and you lost,' laughed Javier as he rode up alongside me.

'What do you mean too scared? Nothing of the sort! I'd race you any day and leave you in the dust,' I replied, remembering gratefully that he wasn't with us at *Las Nutrias* the day I raced Gecko and lost tragically.

'Race me? We'll see about that,' and he shot off in a mad gallop.

Off we went again. Determined not to lose this time, I kicked on Por Si Acaso faster and faster - and boy, did I make that horse run. I'd never had any need to make him run like this before, which meant I didn't fully realise how gutsy and phenomenally fast he was capable of going. With his hooves barely touching the sand, I promptly caught up with Javier. Now side by side, Javier shouted something over at me, but with the thunder of the galloping hooves and the wind rushing past my ears, I couldn't hear a thing. All I heard was Javier laughing and with this, I gave my super steed one last push and he shot off, leaving Javier behind, just as I said we would. We stopped at the top of the incline and I let my poor old boy catch his breath – enough now. As he stood panting heavily, I looked down at the others below me, let go of my reins, stuck my arms out and let out an empowered 'woo!' If someone yelled, the boys could never resist the urge to reply, so my call was echoed by a variety of sounds from everyone down below.

Since Tío Flaco was stuck in town, I would accompany Cabeza to check the water supply at *La Vigilancia* on weekends. During the week this was Chaque's job, but as he would always go home at the weekends, checking the motors became our obligation and as the days got cooler and shorter, the weekend work became much lighter. I only really went as company and gate opener, but I enjoyed our water pump gallivants on Saturday and Sunday mornings and it was on these trips that we really connected. As the foreman, Cabeza needed to maintain a strict sort of presence, but when it was just the two of us, he behaved differently. We joked about stuff and he opened up more, speaking to me about his life and previous jobs, all of which were anecdotes I loved listening to. Likewise I told him all sorts of stories about my life back home, stories of antics from my school days and the various mischiefs at university, jokingly telling him I was good at shooting because my friends and I would drive around town and shoot Freshers with nerf guns. These sorts of stories would always make him laugh and he'd always end by saying, *'sos loca vos.'*

However, the tone of our conversations over the final two weekends shifted to reflect the fact that I was leaving. I went round and round about the mixed feelings I had of wanting to be reunited with my parents and be there to know they were safe, whilst simultaneously being desperate to stay at *San Eduardo*.

'Then why the hell are you going? You're better off staying here,' he would say as we discussed the subject. He did understand, but deep down he was sad to see me leave, too. Strangely however, he did say one thing which made me feel better:

'If you're so determined to come back, you mustn't feel so sad about leaving.'

I thought about this a lot, and he was right. Thanks to this, he helped put my departure into a new perspective which lightened the burden a little, although the sadness was still evident. One afternoon, he drove around to look at some fencing that had been reported as having come down so I went with him on the jolly. I had my phone with me, which we connected up to his truck's Bluetooth to listen to some different music. We listened to some REM and Queen, both of which he enjoyed, and also some Spanish music. 'You can complain all you want about our cuisine, but you can't deny that our music is amazing,' I would tease him.

As I developed a taste for their music, I had added songs to my playlists and, along the way, one of these songs came up. However, it wasn't happy *cuarteto*, but rather a very sad song called '*Ahora Que Te Vas*' (Now that you're leaving) by a singer called Christian Daniel. '*Today I wake up with my soul in pieces, Today I say farewell without wanting to say goodbye'.* As the opening lines were sung, Cabeza exclaimed

'I love this song!' and he cranked up the volume.

I was feeling too glum to listen to this very sad song to plunge even further into gloominess, so I tried to persuade Cabeza to skip the song, but he wasn't having it.

'I'm going to play this when you leave.'

'Don't you dare,' I responded in an abrupt but jokey tone, although it was meant in every way but a joke.

We listened to it once and when it finished, we played it again. This time, the volume went up even further and we both ended up singing along loudly. I was determined not to shed any tears about leaving (more than anything because I had actually made a bet with Pato, who said I would), but I could feel myself fighting watery eyes as we sang together and I listened closely to the words. The timing, however, was excellent and the song finished just as we approached the *casco*. We finished the journey with this song and drove the truck into the barn in silence. Cabeza parked up, switched the engine off and we looked over at one another with an expression that said, 'We don't speak of this to anyone,' without actually saying it in words. As we exchanged this look, I noticed his eyes were shiny and watery. We left the vehicle and as we walked off, I nonchalantly said to him, 'I'll boil the kettle.'

Regardless of Cabeza's encouraging words, I felt completely different this final weekend, the polar opposite of how I had felt throughout my time there. I felt heavy and totally listless, unenergetic, with no enthusiasm to do anything. The depressed side of me just wanted to sit alone in silence, whereas the other part of me was saying I should do everything, as it would be my last time. Thankfully, this attitude fought harder and on Saturday afternoon, I went to the feedlot with Cabeza, Paisa, Pato and Pirulo to tag and vaccinate calves. At first I was on vaccinating duty, but my sluggishness made me resign my position to Paisa and I instead took Pato's position at the gate, with not much avail either. The gates, which usually I could slide open and shut with no problem, now felt heavy and difficult to glide across the rail. I genuinely felt as if I couldn't do anything and that's a feeling I truly hate. In the end, I just sat on the ground out of their way, helpless and weak whilst my friends worked on.

Pirulo glanced over at me and then turned to the others. 'It breaks my heart seeing her so sad,' I heard him say, and the others all grunted in agreement.

Whilst I was sat moping, I realised something which hadn't really struck me until now. I was sleeping about 6 hours each night, sometimes less, and easily managed to work through the day. I never slept the *siesta* like the others did and I was always moving around doing something, but I never felt tired. I didn't take a single day off during the three months – I only slept in two mornings on a weekend. That was all, yet I was full of energy and I now realised that it was because everything about that place was so happy and full of life. Of course, some days were without a doubt physically exhausting, such as working in the cattle pens doing the PD testing, but when your work involves doing something you absolutely adore and you're surrounded by laughter and good people, you soon learn that, quite simply, *that* is the only stimulation you need to keep going. Head held high and glowing with energy. Some nights, I got into bed and would be asleep before my head touched the pillow, which is absolutely not normal for me. The next morning, however, my alarm would go off and I would leap out of bed and open my door into the early morning darkness with no feeling of laziness or unwillingness. Each and every day was an exciting one and I was always anxious to know what each new day would bring me.

Despite being overcast in the morning, by lunchtime it had turned into a beautiful, sunny day. After eating, I sat at the table on the porch of the *matera* with Pato and Pirulo before they went off for their *siestas* when Ricardo came over, holding a small brown paper bag in one hand and a polythene bag in the other.

'Sofi,' he called me, both looking and sounding timid. 'I have something for you, a small leaving gift.'

I jumped up off the bench and approached him, feeling somewhat shy and embarrassed that he was another person who had something for me that I would be able to give nothing in return. Firstly he passed me the polythene bag and inside was a lovely tanned leather waistcoat, with metal buttons made to look like shotgun rims. I gasped as I pulled it out of the bag and held it up to have a proper look at it.

'I've had it a long time but it doesn't really suit nor fit me. Then, when I saw you, I thought it'd be perfect for you.'

'It's beautiful! I love it.'

'Hang on, there's something else,' he added quickly and outstretched his arm holding the little paper bag. I held out my hand to receive it, but before handing it to me he cautioned; 'Careful, it's fragile.'

Having received the warning I took it gently from him and liberated the spherical item from its protective wrapping. Given its shape and the metal straw poking out from the top of the bag, I had a fairly good idea of what it might be, but even so I was surprised when I stripped back the last piece of wrapping and in my hand sat a clay *bombilla* with '*Caichué*' engraved on it. I mentioned at the very beginning that the traditional name of the ranch was the native Indian word for 'sweet water' - and this was it.

'My sister made it. She's really into pottery and it's made entirely with clay from *La Pampa*,' he added. I turned the *bombilla* upside down and saw on the bottom the engraving of his sister's initials.

I placed it down carefully on the table out of harm's way and gave Ricardo a massive hug. He was such a sweet man and I was going to miss him. Once we separated he promptly trotted off looking busy. As I retook my spot on the bench admiring the beautiful *bombilla*, I simply couldn't get over how deeply touched I was by all these special gifts. I was quite simply gobsmacked and what's more is, later on that same day, I was caught by Sergio's wife Lu who, as someone who works in textiles, had knitted me a little *Pampa* scene with the sun rising into the blue sky above the sand dunes.

'I shall hang it opposite my bed so that every morning I feel like I'm waking up in *Etcheto!*' and I dropped down and gave their little boy, who had been sent to deliver my present to me, a big cuddle before getting up and giving an equally big hug to Lu and Sergio.

Sunday. This was it - my last full day. Miguel thought it'd be a nice idea to spend it fishing, so just as we usually would, we loaded the trucks with everything and went at around midday after I had helped Pato, Pirulo and Chango feed the calves. Pato and I moved the calves into the pens to munch on the bales of alfalfa whilst Chango helped Pirulo by opening gates to let the tractor in to dispense the hard feed. Pato told me to go and lie on one of the bales so that I would be surrounded by the calves eating and, totally proving me wrong as I thought they wouldn't come near me, shortly after I lay still, I was indeed surrounded by calves nibbling away, totally careless of me being so close to them.

As we returned to the *casco* to load up all the fishing gear, we walked past Pato's accommodation on the way.

'Hang on,' he said to me and he quickly slipped inside.
In a few moments he came back out and handed me a beautiful knife; the handle made out of antler and wood and the leather sheath stitched with white leather thread.

'I want you to have it,' he said as he shyly handed it over to me.

'It's beautiful, thank you. Did you make it?' He nodded, and I thanked him again for his beautiful gift with a huge hug.

As always, our day at the lagoon amongst the dunes fishing was a relaxing and pleasant one. We had some snack food throughout the day but decided not to have a proper dinner. Instead, a fire was lit and Sergio came with his wife and their little boy with some home-made cakes and Ricardo with some *vizcacha in escabeche*, rolls of bread, wild boar salami and dry-cured venison hams in olive oil - both of the latter he had made himself. We sat by the fire watching the sun go down behind the dunes and chatted happily, reminiscing over my incredible three months there until it got completely dark and it was time to head back.

I went to the *matera* to sit with the boys after having had my shower and they were cooking up the literal bucket-load of fish that we had caught. Several big fish were set aside for the family and Juli to eat, whilst we had the rest. Some were cooked on the *parrilla*, whilst others were gently fried in oil and it was all delicious. There is nothing better than eating your very own catch of the day. There was so much fish that Gecko had come up with his wife and daughter to share it with us and it was then that I realised this would be the last time I'd see him as he wouldn't be coming up to the *casco* for anything tomorrow.

As everyone went to bed I did the same, and this meant saying my first goodbye. I gave Gecko's wife a hug and thanked her for always having some lovely *mate* and *tortas fritas* ready for us every time we went down to *Las Nutrias* to work. I then hugged Gecko and thanked him for everything. When we worked at *Las Nutrias* I would often ride with him, so I got to know him quite well as we got the chance to chat and I told him I was going to miss our talks. We could laugh or speak seriously and I put a lot of trust in him because of this. In fact, at the start of the pandemic, when I was worried about my friends and family back home, he was the one who gave me the most reassurance and I appreciated it greatly. Gecko would always ride along quietly, listening to everything I had to say and would give helpful, reassuring advice, which is what made me confide in him so deeply. As much as he would tease (and trust me, he'd often irritate me hugely with his jokes), his sensible side shone through and he always looked out for me.

I hate goodbyes so after this, I quickly rushed to my room to avoid the lingering sadness. I saw Juli in the kitchen labouring away, so I sat with her a little while, just to chat. She loved chatting and was very good at it. You'd never be lost for things to talk about with her and right now, I needed something like this and although I didn't stay long, I made the most of her company for a while. It got late and pretty much everyone else had gone to bed, but when I tried to close my eyes and get to sleep, I simply couldn't. So, instead, I put some jumpers on and went back outside. What for exactly? I didn't know, but the night was fresh, the sky was clear and the bright full moon shone over the entire place. Without needing a torch due to the immense moonlight, I walked out of the entrance to the *casco*, through the gate with the wagon wheels and carried on up the track, stopping at the cattle grid as I didn't fancy a confrontation with the families of buffalo. I sat under one of the poplars and just looked around and listened to the unique sound of the *Pampa* night-time. How long I was there I don't know, but after the very last lights at the *casco* were switched off by the last people up, I figured it must have been late enough and wandered back to try and get to sleep.

FAREWELL, *SAN EDUARDO*

As I got dressed, pulled back the curtains and stepped outside, I felt totally hollow inside but just like any morning, I walked over to the *matera* and sat with the boys, waiting to greet the others on their return from town.

I had to pack everything and sort myself out so I made the very hard decision to not go out to work, not even for the morning, as it meant I would be too pressed for time. Miguel had planned for us to leave after lunch and I wanted to make sure I had time to do everything peacefully rather than in a hurry. I loitered briefly with El Tío whilst he tacked up his horse and once he set off I went to sort out my room, which luckily didn't take very long due to the very few items I had brought with me. With everything quickly packed, I went for a walk around the *casco* and spent a little bit of time sitting with Luna the dog, another member of the ranch that I would miss a lot. I returned to my room, having nothing else to do, when I heard on the radio that Pato, Pavo and Enano, who were working in *Don Juan*, would soon be ready to be picked up and were calling for Pirulo to go and get them. I quickly rushed out to meet Pirulo before he drove off so I could go with him at least, to give me a little something to do and give me some extra time with my friends.

Lunch was not the usual. I had asked Miguel if he wouldn't mind me having my last meal with the gauchos so I could make the most of my last day with them, to which he said that it was no problem and perfectly understandable. He and Patricia had both noticed the strong friendship I had developed with them over the short time I had been there. Everyone was quiet, not really chatty and certainly not up for making jokes. One of them didn't even come to lunch and when questioned why, he replied with a long face saying he simply wasn't hungry, which broke my heart. Not really knowing what to do afterwards, everybody went to their rooms as usual. Since they didn't feel like talking, there was no point in staying to chat at the *matera* because nobody, myself included, had much to say. I wandered around the *casco* again and as I went to see one of the horses which had been left in the round pen in front of Pirulo, Javier and Enano's room, when they saw me they called me inside to sit with them. 'Don't be alone,' they said. 'Come and sit with us.'

They put on some music and shared *mate* as I watched them working with leather. Enano had made good progress on his new lasso and Pirulo diligently continued with his plaited bridle, which I made him promise I would receive a photo of as soon as it was finished. Javier just sat on his bed, joining in on the conversations and snacking on the biscuits and boar pancetta that we were all munching on. As the clock ticked past I became more and more anxious, constantly looking over at the *matera* to see when everyone would be gathering before work. My attention was then diverted when Pirulo stood up, took his *culera* off his shelves and handed it to me.

'I meant to give this to you sooner, but I wanted to have something to give you when you left.'

298

No, I couldn't accept it! He'd already gifted me my *faja* and when he gave me a pair of brand-new *alpargatas* that I told him were too big for me he made me put them on in front of him to prove it.

'I have nothing to give you, Pirulo. I'm sorry, I can't accept anything more from you.'

'Of course you'll accept it. You can't have the *faja* without the *culera* to go on top! If you don't accept it then once you leave here, you'll never hear from me again,' he threatened.

With no other choice, I accepted yet another beautiful gift and gave him a huge hug. Something one of them once said to me will be stuck in my mind forever, which was '*I know what it's like to live with nothing, but even during those days I was happy. As a result, possessions mean nothing to me so if you want something or I want to give you something, then I will give it to you. There is a lot more pleasure in making a gift than having a lot of possessions.*' Ever since those words were spoken, they have been inscribed in my memory. It had also been a wake-up call that, although everyone on the ranch was happy, had a job and a home, not everyone had a happy past. One afternoon, I found myself chatting to one of the contracted workers (not one of the horsemen) and he told me that in his youth, there was a stage when he had to leave home to find work. In the process, he slept on cardboard and newspapers on the streets. However, he persevered and, being an immensely hard worker with an incredible eagerness to learn, he climbed his way up and as a result was the boss of his own small team on the ranch.

'Damn, we're going to miss you *Loca*' said Javier, quietly.

'Me too, I can't believe I have to go.'

'Just remember, it's not too late for you to marry me and stay here. The offer still stands,' added Pirulo, to which we laughed as we remembered that lunchtime from my early days when they all started arguing over who would marry me so I could stay in Argentina.

Not long after, I saw El Tío enter the *matera* so I decided now was a good time to pack my *culera* and quickly put away my *recado*, which I had left at the horse lines. I put my *culera* in my room and then went into the house to get the keys to the barn for my *recado* when suddenly I heard Pavo calling my name. He told me to hurry, as they were all going back to work and they wanted to say goodbye. I rushed out in a fluster, leaving the keys behind and my *recado* on the ground of the patio to deal with later. All the boys had suddenly left their rooms and were getting straight into the trucks without even having their *mates*. Knowing he would be truthful, I asked Pavo why they were leaving so fast, to which he said they wanted to make sure they had gone before I even started loading up the car because they didn't want to be around to see me leave. The moment I had been dreading had finally come around and, in a rush, I gave each and every one of them an enormous hug before watching them drive away. In two shakes of a lamb's tail, I was left standing alone between where the two trucks had been seconds earlier. Just like that they were gone, and I had no idea when I would see them all again. Then it was goodbye to Chaque, who had appeared slightly after everybody else and was off to do his water pumps and motors. It was far from the farewell I had hoped for, but at least I managed to see everyone.

Time moved weirdly slowly from there onwards. I put away my *recado,* Miguel brought the car around and I began taking my bags out of my room to be loaded up. As I walked by with my bags I heard Patricia call me from within the house.

'Wait! Come,' she requested, so I dropped what I was holding and went inside. She came out from another room holding a little clay armadillo, which when turned upside down was hollowed out like a little cup or *bombilla* and its tail served as a handle.

'I'd like you to have this,' she said in her sweet voice as she passed it to me. 'It's very dear to me but I hope that you will keep it forever to remember us by.'

As if I would ever forget them! I admired the adorable little object, which became more adorable as she explained how much she loved it and how she hoped I would always remember *La Pampa* upon looking at one of its treasured armadillos. I felt terrible walking away with something she loved so much, but Patricia absolutely insisted that it come back to England with me. She really was one of the sweetest people you could imagine and when I went back to my luggage, I made sure to wrap it up extremely well with the *mate* I had got from Ricardo, to ensure they both had a safe journey.

Chango, whom I was told hated goodbyes, slipped away quickly to cut up wood around the back of the tractor shed. Cabeza and Pirulo both returned to the *casco*, having dropped off the boys at their horses and sat at the *matera* just waiting, not doing much. I perched with them briefly whilst the rest of the car was being loaded up and as I sat there, Sergio showed up with his little boy in his arms, accompanied by Lu. I've never been particularly good with children but dare I say, this little fellow made my departure a lot easier. Whilst with Sergio, he was making tractor noises (he absolutely loved tractors) and looking around for it.

'Sofi will take you to see the tractor,' Sergio told him, as he lowered him to the ground and walked him towards me.

I picked him up, gladly but hesitantly, and took him to see the tractor, where I opened the door to the cab and let him sit for a while in the driver's seat before he got bored and wanted to get back to Daddy. I hadn't really bonded with the little fellow during all my time there, as he was quite shy, but that afternoon he opened up to me and stood next to me the whole time, holding on to my trousers if I ever let go of his hand.

The time continued to pass slowly and eventually, Miguel, Patricia, Lucas and I were joined on the grass courtyard outside my room by Juli, Cabeza, Pirulo, Brandizi, Ricardo and Sergio with his family. Even all the dogs joined us. Whilst I had Sergio's little boy glued to me on one side, Luna was on the other, lying on her back so I could scratch her tummy.

'It looks like it'll be more than just the boys that will miss you. Perhaps you should tell your mother you won't be returning alone. You may scare her when you tell her that, but I think she'll laugh when she sees you walk off the plane with a little black dog,' giggled Patricia.

'I wish I could,' I replied to her, remembering all those mornings I opened the door and she would be there waiting for me. We dragged the process out for as long as we could but inevitably, the time to depart arrived, so I said my final goodbyes. As I went around hugging everyone there was no way I was able to thank them all enough for all they each did for me. I struggled to keep myself together as I hugged dear Juli goodbye, who was quietly stood by herself trying to hide her red face and watery eyes.

'Look at us, eh? The two who said we didn't want to be here when she left and ironically, here we are,' Pirulo scoffed to Cabeza. It was true. When the three of us had spoken, they were adamant that they wouldn't be there when I left because they didn't want to say goodbye, yet there they were and I appreciated that.

I gave Luna a final scratch on her tummy, placed my hand on Sergio's son's head and then as I walked away, I heard a little voice.

'No, tractor!' cried the little boy.

'Don't worry, I promise I will be back soon and we will go and see the tractor again, OK?' I knelt down and gave him one last cuddle.

The car doors shut and as we drove away, I refused to look back. I was absolutely gutted but I didn't drive off in a stream of tears like everyone had anticipated I would. No, in some weird way the pure determination that I would return, the fact that I promised everyone I would, and Cabeza's words kept me strong: 'If you're so determined to come back then you mustn't feel so sad about leaving,' I recalled in my mind over and over again.

As we drove down the long track across the ranch to the entrance I looked at everything one last time, taking in this picturesque and interminable paradise until we got to the final gate. We went through, turned onto the tarmac and were off, on our way back to *Buenos Aires*.

THE SILENT CITY

Due to the restrictions imposed by the government, I travelled with no less than 10 pages' worth of documents from the Argentinian government, the British Embassy and my flight tickets, all of which granted me permission to travel within the country, should we have been stopped by the police. The rules were brutally strict, one of them being that I could only enter *Buenos Aires* 24 hours before my flight and I could only travel directly from my accommodation to the airport.

The long roads were empty but what shocked me most was the arrival in the city. Given how late we left the ranch, it was almost two in the morning by the time we arrived and since I was staying at Miguel and Patricia's house overnight, on our way, I had to collect the suitcase I had left at Leonardo's house. As we drove through the city, it was actually frightening. This city, which last time I was there was full of life and happy people now lay dead; not a soul or car on the roads anywhere, just the occasional police vehicle here and there. This virus now suddenly felt terribly real and I began to feel scared. I had seen pictures online of all the major cities totally empty and carless and heard stories about how the canals in Venice became so clear that dolphins swam in them and in Nairobi, lions sunbathed in the middle of the streets. Having been sheltered away in my little corner of paradise at *San Eduardo* for so long with no contact with the outside world meant seeing this gorgeous city so quiet and lifeless was very difficult to take in.

It was hard saying goodbye to Miguel and Patricia. They were both so lovely and had been so amazing to me. I would miss Miguel's sense of humour and Patricia's all-round kindness. Just before walking out the door, Patricia went into her storage cupboard and grabbed a bag of *mate*.

'For you; to enjoy when you miss us back home.'

I told them the doors in England would always be open to them, the same way they had opened their doors and hearts to me.

The taxi journey to the airport the following morning was no easier than the journey into the city the night before. I sat in the back seat, mask on and with a transparent plastic sheet taped between the front and back seats of the cab, separating me from the driver. I gazed out of the window at the silent city and it was even weirder during the day not seeing anyone. As we stopped at the traffic lights on the *Avenida 9 de Julio* directly in front of the *Teatro Colón*, we were the only vehicle on that entire stretch of road. At the airport, there was only one entrance where two men in masks and white Hazmat suits stopped me and took my temperature with a laser. All clear, I went on through to leave my luggage and endure the 15-hour flight where everyone had to wear their mask the entire time. You could only stand up to go to the loo and since it was a specially chartered repatriation flight, all we were served throughout the whole journey was one cup of water and half a ham and cheese sandwich. The plane was cramped and uncomfortable, with every seat occupied by people like myself who had become stuck in

Argentina due to COVID and were on their way home to safety. However, I couldn't help but see the irony in this though. I was going to be in a room for two weeks isolating and then bound by government restrictions forcing me to stay at home for as long as they decided. I couldn't help but feel that I was leaving freedom and safety for confinement, not the other way round. I would surely be more stuck and at risk of this virus in England than I ever had been on the ranch.

Gatwick - empty. There was minimal staff around and none were wearing masks, gloves, or any kind of protective gear. My flight was the only arrival at the airport. I collected my case and, being so hungry from the flight, bought a Twix bar from a vending machine, which was all they had. As I exited I called my father, who told me to meet him at the car park and as I got out of the lift and stepped onto the pavement outside, I saw him on the opposite side of the road next to the car, masked up and wearing gloves.

'Hello,' I said to him casually, to which he replied with a 'hello' back.

He crossed the road and, as he stepped towards me to take my case, I took a step back to maintain the distance. It was all so cold and emotionless. I couldn't hug him or touch him and I walked to the car with him in silence, as if I had just been with him the day before, not 8 months away from home across the world. I had only just arrived and I already hated what the world had become due to this invisible enemy. He sprayed my hands with disinfectant and opened the door for me, where I climbed into the back seat and saw all sorts of tablets and supplements. My father, being the good doctor that he is, had been researching the virus right from the beginning and was taking into consideration all the new information available that would serve as a good precautionary measure in case I had caught the virus on the plane.

Both with masks on and the windows down, we drove home along the empty motorway. I looked out of the window at the familiar green countryside and up to the grey sky. Being back suddenly felt as if I had never left at all, which hurt me deeply. All those experiences and stories, both from Australia and Argentina, suddenly felt as if they were all not true. It was as if the reality had suddenly disappeared the moment I set foot off that plane. Even on the journey I barely spoke of my trip; instead, the conversation revolved around the bloody plague and I picked my father's brains for everything he knew about it so far. A couple of hours later, we rocked up on the drive where the wooden gates to the house had been decorated in chalk by my mother with a massive 'Welcome Home' and drawings of bunting, balloons and party poppers. Even the neighbours had done the same, which I thought was very sweet. As the gates opened I saw what I could of my mother's face behind her mask filled with pure joy, which hurt deeply as I wouldn't be able to hug her either, not to mention my three dogs who were stood at the garden gate, howling and besides themselves with happiness, unable to believe their eyes that I had finally come back and was standing right there in front of them.

As I entered the house, my clothes went straight into the washing machine and I to the shower. Afterwards, as I sat on my bed staring out of the window, I began to cry and just couldn't stop. All the tears I hadn't shed at the time of leaving the ranch just flowed and I sat there for ages, just staring out of the window and sobbing as I heard my phone pinging with messages all from my gauchos. 'Where are you?' 'How are you?' 'Are you home yet?' With a deep breath, I eventually managed to calm myself down, put on my mask and walked out into the garden, where my parents and the dogs greeted me cheerfully. At last, washed and in clean clothes, I was permitted to give the dogs a hello, although I would still have to wait two weeks before giving my parents theirs.

Eventually, the two long weeks passed, but during that time I stayed in my room, totally isolated from my parents. I was given a few provisions like a kettle, coffee, my *mate* and a small fridge for milk and water, but everything else was brought up to me on a tray, delivered by either one of my parents, who would come up with a mask and gloves on and leave the tray on the stairs outside my room. With very little known about this new virus, they couldn't be too careful. I spent all the time I could on video calls with my gauchos every day, sometimes even having *mate* with them in the morning. When they weren't around because they were out working, I spent my two weeks watching films on Netflix whilst I ate, playing Solitaire and *Truco* (I got the game on my phone) and downloading all my photos. It might not surprise you when I say I even started writing this book, which gave me a lot of peace. It's as if by re-telling my experience, I have been able to relive it all.

A YEAR ON

Although the world is hesitantly trying to get back to normal, the truth is that it's still far from it. When I left *San Eduardo*, I had hoped that I could return the following year, make up some money working back in England and then go back. But, this wasn't the case.

However, through thick and thin, the gauchos and I have stayed in touch. I maintain contact with them all, and they with me, and I'll speak to at least one of them on the phone every week. As soon as I finished my quarantine, I found work doing a harvest and when I was up hauling grain at 1am I was fortunate to have a gang of cheerful, noisy people four hours behind me to keep my spirits up and keep me awake. Occasionally, I'll catch someone in a group and I'll be able to speak to a few of them at a time. A couple of times Ruso has called me so I can see and speak to Luna, who for a while would look around when I called her name and she would jump up on him when she couldn't find me. He even told me that for the first few days after I left, whenever he went to see Juli in the kitchen, he would find Luna sitting outside my door.

Soon after I left, Gecko and Chaque each started breaking in a filly; a buckskin and a chestnut respectively. 'I'm making her really good and calm so she can be yours to work when you come back,' they both said. Meanwhile, I got pictures and videos of the progress each made with their respective projects. It is safe to say, calm is an understatement after receiving a video of Chaque on his little chestnut riding her around with nothing more than a rope around her neck.

Other times, I would call one of them and catch them during their weekend off in town with their friends. 'It's Sofi,' they would tell their friends, who would then all gather around the camera saying 'ooh, the English girl from the ranch!' One night, I was messaging Pirulo and he was at a party with some friends, singing karaoke and playing the guitar. *'If you call me now, the chicos said they will sing for you'* he wrote. Without hesitating, I rang him immediately and on my request with their acoustic guitar, they sang any song I asked for, along with others I hadn't heard of. Speaking of Pirulo, he sent me a photo of his finished plaited bridle, just as he said he would. He didn't forget. He also wrote down all the rules of *truco* in a clear, easy-to-follow way, which he sent to me to look at and study, in the hope that one day I might improve my game. Also, he and Enano would frequently take in orphaned creatures, bottle-rear them and as they became tame and friendly, they would even have them in their room with them. They would send me videos of them having *mate* and working with leather and there'd be this orphaned baby animal, of whatever species, lying on the floor between them like a dog. The year I left it was a red Angus heifer calf and the following year it was a week-old red deer doe. When I asked them what they had called them, they told me both had been named Sophia.

What really makes me smile are the occasional videos I get from Sergio of his little boy, where when they mention my name he will look up at them and say 'countryside?' followed by 'tractor' with him making tractor noises. This is something I am always really touched by because I didn't expect a one-year-old to have any recollection of me at all, especially since I never really interacted with him much in the first place. I did tell him before I got in the car to leave that I would take him to see the tractor when I returned and it seems he's making sure I don't forget my promise.

Miguel and his family have continued to be so sweet. They've said I am welcome back anytime and whenever they go to *San Eduardo*, I will often get photos from Patricia. I do hope to see them again soon, whether it's me going back or them coming over to England.

Everyone always keeps me well informed of how everything is going, what they are up to and whenever Claudio the vet is there he gets one of the boys to send me his regards. Sadly, a couple of the boys have moved on to other ranches, but this hasn't really changed anything. We have stayed in touch and instead they tell me about the work they are doing there, how it's different to *San Eduardo* and they've introduced me to some of their new co-workers via video call. I've also been introduced to some of the new chaps at *San Eduardo*, and I even helped one of them with his English homework for a couple of months whilst he completed school. I love this, as it means I've been able to feel connected to the place, despite being almost 8,000 miles away. I still get sad, though. I'm frequently asked when I'm coming back and rather uselessly, I can only ever answer, 'I don't know.' I miss them and I do think of them every single day. There have even been some nights that I've dreamed the exact scenarios, exactly how they were. A few months after leaving I dreamt the entire sequence of the day Javier and I moved all those

311

cattle, just the two of us, when I was riding Gusano and the radios were out of charge from the storm the night before. Reliving something I've done in real life and then dreaming about it is something I've never experienced before. On a separate occasion, I fell asleep to the sound of the neighbour's cows lowing in the night and it reminded me of when I fell asleep to the sound of the deer and the calves in the feedlot. Anyway, I must have fallen asleep with my brain thinking I was there and when I woke up in the night to get a drink of water from the bathroom, I stumbled, half asleep, to what I thought was the bathroom but instead, walked into my bedroom wall. Subconsciously, I had headed in the direction of the *en suite* in my bedroom at the ranch. This explanation was received with strange expressions the following morning when I was asked by my parents, 'Did you hear that massive thud last night?'

At *San* Eduardo, for the first time ever I was speaking only Spanish 24 hours a day and as a result my English became slow. I even started dreaming in Spanish, something I had never done before, but the thing I found funniest was that in just 3 months my accent changed. All my life I've been told I speak Spanish with a slight English twang and people like Miguel and Claudio picked up on this. The gauchos on the other hand said I spoke very perfect Spanish, with no accent at all. However, when I went to Spain to visit family they all said I now sound Argentinian and what's more, when I spoke to people I didn't know, the majority thought I *was* Argentinian - to the point that one person asked me where abouts in Argentina I was from.

As for the pandemic, Argentina had its highs and lows just like everywhere else but generally did better than England, which was surprising considering everyone thought the Argentinians would drop like flies because of how they shared *mate* and beer with anyone and everyone. Things opened up much sooner and small gatherings were permitted within a few weeks of me leaving. Inevitably, some cases did enter *La Pampa* and a couple of cases did arise at the ranch. In fact, I was informed that one of the gauchos who had contracted it got seriously ill and was on a ventilator in hospital for a few weeks, plus another few weeks on top of that in intensive care, fighting for his life. I prayed he would recover. He was out of work for two months and lost a quarter of his bodyweight whilst the doctors fought to keep him alive but thankfully, he pulled through and made his full recovery at home before returning to the ranch, back to his usual self.

Whoever is reading this, I'd like to just leave you this short message: If you have the opportunity to travel, see the world and meet its people, do so. Whether you're a student on a gap year or you're making some life choices as someone further along the line – do it. If you can, do not hesitate. Immerse yourself in the culture of the world and its people. There may be some uncertain, dangerous places and people, but thankfully there are more good than bad. I include Australia in this statement when I say, the pride of looking back and saying, 'wow, I've been there, seen that and met all these people' is the most valuable feeling one can have. 'Have stories to tell and not stuff to show,' I heard someone say once. The memories will stay with you forever and you will always have your own stories to share. The world is a beautiful place and I cherish the friends I made whilst far away.

In the end, it all comes down to this: I made a promise, and I hate breaking promises. I promised them all that I would someday return and of all the promises I've ever made, it's one of the ones I am most determined to keep.

GLOSSARY OF TERMS

Alfajores: Typical South-American sweet. They are two shortbread-type biscuits, traditionally circular, with condensed milk in the middle then rolled in ground coconut, which sticks to the condensed milk around the middle. They can also be coated in chocolate.

Alpargatas: Espadrille shoes. In this case they are usually fabric, sometimes leather, and have rubber soles (not straw).

Amiga: (Feminine. Masc. = *amigo*). Friend.

Asado: Barbecue. The iron grill the meat is cooked on is referred to as a *parrilla*.

Boina: Like a beret, also known as a Basque hat. Practically a defining item for the gauchos, *boinas* can be made of knitted cotton or felt and come in a wide range of sizes and colours.

Boludo: Stupid, idiot, moron.

Bombachas: Chino-style trousers made from heavy cotton.

Bombilla: The receptacle from which *mate* is drunk. It translates to 'bulb' due to its shape. Although they can be made from any material, traditionally *bombillas* are made from small, dried, hollowed-out squashes.

Buckskin: Horse coat colour caused by a dilution gene in a bay horse. Like bays, buckskins have black manes, tails and black legs up to the knee and hocks, but instead of brown are a creamy-golden colour over the rest of the body. Unlike a dun, buckskins do not have a dorsal stripe.

Buen día: Good morning.

Cabeza: Head. Also used to refer to cows which were served in the first month (of 3) of being with the bull.

Caldén: (Pl. *caldenes*): Prickly Acacia (*Prosopis caldenia*) in abundance and deemed an icon in *La Pampa*. Grows both as a bush and as a tree. Although endemic and treasured in Argentina, the Queensland government in Australia have clearance programmes in force to eliminate it.

Carancho: Crested caracara – a type of falcon (*Caracara plancus*).

Casco: The ranch's main homestead.

Chancho: Wild boar.

Che: A slang word without real meaning, sometimes used as "mate" or to get someone's attention (like 'hey' or 'oi').

Chicharrón: The solid fat left over after the tallow has been removed from rendered down fat, usually beef fat. The gauchos loved eating this on bread, either alone or with a sort of *chimichurri* sauce.

Chicos: (Masculine. Fem. = *chicas*). Boys (in the tone of 'chaps' or 'lads').

Chinchulín: Beef small intestine.

Chuzo: The Spanish word for icicle, used to refer to a stag with antlers that end in a long, speared point at the tip rather than with a crown.

Cola: Tail. Also used to refer to cows which had been served in the third month (of 3) of being with the bull.

Colt: A young male horse (usually uncastrated and under 4 years old).

Cotorra: Monk parakeet (*Myiopsitta monachus*). A small, green parakeet found in abundance around *La Pampa* and make nests out of *caldén* twigs in windmills.

Criollo: Creole – Latin American people of Spanish descent but in Argentina, this is also the name of the native cattle and the Criollo horses which are small, incredibly sturdy horses often characterised by stripy markings up to the knee.

Cuarteto: A genre of music from the province of *Córdoba*. *Cuarteto* songs are usually upbeat and lively and common instruments include the keyboard, drums, accordion and shaker, like a maraca, to sustain the beat.

Cuerpo: Body. Also used to refer to cows which had been served in the second month (of 3) of being with the bull.

Culera: A leather belt worn over a *faja*. Purely decorative, *culeras* are not used to uphold trousers.

Cumbia: Another genre of music from *Córdoba*. *Cumbia* is a slower, gentler style of music compared to *cuarteto,* mostly revolving around the use of maracas, flutes and light percussion (eg. timba drums).

Drafting: Separating out cattle into desired groups (eg. age, weight, sex etc.)

Dun: Horse coat colour. Much like a buckskin, duns have black manes, tails and black legs up to the knees and hocks. They are also a creamy-golden colour but duns are characterised by a black dorsal stripe which runs across the length of the horse's back, from the base of the neck to the top of the tail.

Empanadas: Similar to a Cornish pasty, this traditional pastry is usually filled with minced meat, onions, olives and cut up boiled eggs and then deep fried.

Escabeche: (Pronounced: Es-ka-BECH-ay). A vinegar-based sauce with herbs and spices used to marinate and preserve meat or fish.

Faja: A long fabric (usually woollen) belt which is wrapped around the top of trousers, similar to a cummerbund.

Forelock: A horse's fringe.

Gaiters: Equestrian clothing item, most commonly made from leather, which you wrap around your lower leg to prevent pinching from stirrup leathers.

Gelding: A castrated male horse.

***Guarda Pampa*:** The name given to the traditional Argentine pattern, consisting of a chain of diamonds, which is frequently associated with polo equipment in the UK (belts, girths, horse headcollars etc.) Believed to be an insignia created by the Mapuche Indians around 5,000 years ago its symmetry, colour combinations and the number of points on the diamonds all represent different things.

***Guarda Sofi!*:** Careful Sofi!

Handling system: What is used to work cattle in order to filter them and restrain them in order to undertake operations, eg. tagging, vaccinating, PD testing etc. The cattle enter the system through a bottleneck, where they pass down a *race* or *chute* before reaching the crush, which is at the end and contains the yoke and other restraints to stop the cattle from moving.

Hectare: Area unit of measurement. 1 hectare = 2.471 acres. Therefore, *San Eduardo* at 52,930 hectares = 130,790 acres.

Heifer: A cow that has not yet undergone her first calving.

Hocks: The 'knee' on a quadruped's back legs which is angled backwards.

Hogging: Cutting off a horse's mane, usually leaving a short, prickly brush along the crest of the neck.

***Jineteada*:** Argentine bucking bronco where riders compete for cash prizes (and street credit). Different categories require you to stay on the horse for different periods of time (between 8 and 12 seconds) and of those who manage to stay on the horse, the best performance wins.

***Jote*:** Black vulture (*Coragyps atratus*).

***Loca*:** (Feminine. Masc. = *loco*). Mad, crazy. Can also be used more affectionately like 'bonkers'.

Maned wolf: A large canine of South America, the maned wolf (*Chrysocyon brachyurus*) is neither a wolf nor a fox. In Argentina it's known as an a*guará guayzú*.

Mara: The fourth largest rodent in the world (*Dolichotis patagonum*). A funny looking creature that looks like a cross between a small deer and a rabbit.

Matambre: Cut of beef. Similar to *vacío*, these cuts are essentially flank and skirt steak.

Mate: (Pronounced 'ma-teh'). The Argentinian version of tea which is drunk by pouring hot water onto   loose leaves in a bulb-shaped recipient (*bombilla*) and drunk through a straw. *Mate* is a social drink, as everyone shares it and drinks from the same straw.

Matera: Like 'tearoom', this is the room where you sit and drink your *mate*.

Monte: Used to refer to the areas with thick vegetation and *caldenes*.

Olivillo: A common silvery plant which grows throughout the *La Pampa*. Its name comes from its leaves, which are a similar shape to olive leaves.

Overgirth: A second girth (not used to hold the saddle in place) that is instead used to hold on the extra cushions the rider sits on.

Pampa: Native Indian derived word for 'plain'. Used to describe the open sand dune area of the ranch.

Pañuelo: Handkerchief or neckerchief. In this context, a *pañuelo* is a neckerchief which forms part of the traditional gaucho uniform and worn more like a tie than as a bandana.

Parrilla: The grill for a barbecue/*asado*.

PD Testing: 'Pregnancy-diagnostic' testing; checking whether cows are in calf.

Pejerrey: Fresh water Argentinian silverside fish (*Odontesthes bonariensis*). Loosely translated to 'king fish', this name may have been assigned due to the fish's popularity and high esteem as a result of its flavour and being easy to cook.

Pequeña: (Feminine. Masc. = *pequeño*). Small, small one, little.

Peso: Argentinian currency. When I arrived to Argentina in February 2020, £1 = 76 *pesos* but since the pandemic the currency is ever weakening, reaching £1 = 140 *pesos* in mid-January 2022.

Piquillín: (Pronounced 'picky jean') The little red berries on a particular thorn bush endemic to Argentina (*Condalia microphylla*). Very much like a red currant, these berries were edible and grew on wild bushes across the ranch.

Postero: The person in charge of a *puesto*.

Potrero: A smaller field, like a paddock, used to keep the horses in which usually enclosed water-extracting windmills.

Puesto: Ranch posts. Some ranches are so big you have people living in their own house at a certain location on the ranch looking after that particular area and its cattle. The person who looks after a *puesto* is called a *postero*.

Qué pasó?: What happened?

Quincho: The stand-alone building at the ranch with a thatched roof. In Argentina, Uruguay and Paraguay, a *quincho* is a place that is especially equipped for *asados*, dining, meetings and social gatherings.

Rebenque: A type of whip the gauchos use made of leather. It has a long, hard handle with a strap of loose leather trailing.

Recado: Traditional Argentinian 'saddle'. Comprised of various layers; typically a sponge is used as the 'saddle cloth', with a blanket on top followed by the saddle-type piece of leather (*basto*) which is all held together by the girth and stirrups. For rider comfort, a layer of blankets and sheepskins are tied on top with an overgirth.

Redomón: An untrained or only partly trained horse.

Roseta: A particular type of grass burr. In *La Pampa*, the *rosetas* from one of the types of grass were hard and woody, star-shaped and incredibly sharp, more like thorns than a burr.

Rosillo: Strawberry roan – a horse with light brown hairs mixed in with white to give a uniform pink sort of colour across its body, rather than having each colour in patches.

Salitral: Salt marsh.

Sos buena: You're good (at something).

Sos campera vos: 'You're proper country.' (For masculine, you would change *campera* to *campero*).

Sos mala vos: You're bad, you're mean. Usually meant in a playful, friendly way.

Tira de Asado: Jacob's ladder - one of Argentina's most traditional cuts of meat where an entire rack of ribs is cut in cross-section, rather than each rib cut out.

Tordilla: (Feminine. Masc. = *tordillo*). The diminutive of the word *torda* which is a horse of grey colour.

Tortas fritas: Dough fried in beef fat. Usually salty but can be made sweet by putting condiments on top, ie. jam, '*dulce de leche*' (condensed milk), or sugar.

Truco: A card game, common in Argentina and Uruguay, played with a Spanish deck of cards.

Vizcacha: (Pronounced 'vith-catcha') A large rodent (*Lagostomus maximus*), around the size of a Jack Russell terrier that is similar in appearance to a chinchilla. They live in underground colonies and the males are characterised by a thick, black wiry moustache which crosses their face from cheek to cheek.

Vizcachera: The burrows made by the vizcachas.

Withers: The highest part of a horse's back: the bony bit at the base of the neck between the shoulder blades. A horse's height is measured up to its withers.

Yarará: A highly-venomous pit viper from the *Bothrops* genus. Considered one of South America's most deadly snakes, the *yarará* can reach 2 metres in length and is viviparous (gives birth to live young rather than lay eggs).

Zaino: (The 'z' is pronounced as 'th'). Horse coat colour – dark bay.

Fancy taking a peek at the beautiful landscape and meeting everybody?

Follow
@ridelikeagaucho
on Instagram for photos, videos and updates!

Printed in Poland
by Amazon Fulfillment
Poland Sp. z o.o., Wrocław